A Primer for the Monte Carlo Method

A Primer for the Monte Carlo Method

Ilya M. Sobol'

CRC Press

Boca Raton   Ann Arbor   London   Tokyo

Library of Congress Cataloging-in-Publication Data

Sobol', I. M. (Il'ia Meerovich)
 [Metod Monte-Karlo. English]
 A primer for the Monte Carlo method / Ilya M. Sobol'.
     p.   cm.
 Includes bibliographical references and index.
 ISBN 0-8493-8673-X
 1. Monte Carlo method.   I. Title.
 QA298.S6613   1994
 519.2′82—dc20   93-50716
                                                             CIP

publishing history

This book was published in Russian in 1968, 1972, and 1978. While it is a popular book, it is referred to in rigorous applied papers; teachers also use it as a textbook. With this in mind, the author largely revised the book and published its fourth edition in Russian in 1985.

In English, the book was first published in 1974 by Chicago University Press without the author's permission (the USSR joined the Universal Copyright Convention only in 1973). The second English publication was by Mir Publishers, in 1975 and 1985 (translations of the second and third Russian editions).

The fourth edition of the book was translated only into German (1991, Deutscher Verlag de Wissenschaften).

abstract

The Monte Carlo method is a numerical method of solving mathematical problems by random sampling. As a universal numerical technique, the Monte Carlo method could only have emerged with the appearance of computers. The field of application of the method is expanding with each new computer generation.

This book contains the main schemes of the Monte Carlo method and various examples of how the method can be used in queuing theory, quality and reliability estimations, neutron transport, astrophysics, and numerical analysis.

The principal goal of the book is to show researchers, engineers, and designers in various areas (science, technology, industry, medicine, economics, agriculture, trade, etc.) that they may encounter problems in their respective fields that can be solved by the Monte Carlo method.

The reader is assumed to have only a basic knowledge of elementary calculus. Section 2 presents the concept of random variables in a simple way, which is quite enough for understanding the simplest procedures and applications of the Monte Carlo method.

The fourth revised and enlarged Russian edition (1985; German trans. 1991) can be used as a university textbook for students-nonmathematicians.

preface

The principal goal of this book is to suggest to specialists in various areas that there are problems in their fields that can be solved by the Monte Carlo method.

Many years ago I agreed to deliver two lectures on the Monte Carlo method, at the Department of Computer Technology of the Public University in Moscow. Shortly before the first lecture, I discovered, to my horror, that most of the audience was unfamiliar with probability theory. It was too late to retreat: more than two hundred listeners were eagerly waiting. Accordingly, I hurriedly inserted in the lecture a supplementary part that surveyed the basic concepts of probability. This book's discussion of random variables in Chapter 1 is an outgrowth of that part, and I feel that I must say a few words about it.

Everyone has heard, and most have even used, the words "probability" and "random variable." The intuitive idea of probability (considered as frequency) more or less corresponds to the true meaning of the term. But the layman's notion of a random variable is rather different from the mathematical definition. Therefore, the concept of probability is assumed to be understood, and only the more complicated concept of the random variable is clarified in the first chapter. This explanation cannot replace a course in probability theory: the presentation here is simplified, and no proofs are given. But it does give the reader enough acquaintance with random variables for an understanding of Monte Carlo techniques.

The problems considered in Chapter 2 are fairly simple and have been selected from diverse fields. Of course, they cannot encompass all the areas in which the method can be applied. For example, not a word in this book is devoted to medicine, although the method enables us to calculate radiation doses in X-ray therapy (see Computation of Neutron Transmission Through a Plate in Chapter 2). If we have a program for computing the absorption of radiation in various body tissues, we can select the dosage and direction of irradiation that most efficiently ensures that no harm is done to healthy tissues.

The Russian version of this book is popular, and is often used as a textbook for students-nonmathematicians. To provide greater mathematical depth, the fourth Russian edition includes a new Chapter 3 that is more advanced than the material presented in the preceding editions (which assumed that the reader had only basic knowledge of elementary calculus). The present edition also contains additional information on different techniques for modeling random variables, an approach to quasi-Monte Carlo methods, and a modern program for generating pseudorandom numbers on personal computers.

Finally, I am grateful to Dr. E. Gelbard (Argonne National Laboratory) for encouragement in the writing.

I. Sobol'
Moscow, 1993

introduction

general idea of the method

The Monte Carlo method is a numerical method of solving mathematical problems by the simulation of random variables.

The Origin of the Monte Carlo Method

The generally accepted birth date of the Monte Carlo method is 1949, when an article entitled "The Monte Carlo method" by Metropolis and Ulam[1] appeared. The American mathematicians John von Neumann and Stanislav Ulam are considered its main originators. In the Soviet Union, the first papers on the Monte Carlo method were published in 1955 and 1956 by V. V. Chavchanidze, Yu. A. Shreider and V. S. Vladimirov.

Curiously enough, the theoretical foundation of the method had been known long before the von Neumann–Ulam article was published. Furthermore, well before 1949 certain problems in statistics were sometimes solved by means of random sampling — that is, in fact, by the Monte Carlo method. However, because simulation of random variables by hand is a laborious process, use of the Monte Carlo method as a universal numerical technique became practical only with the advent of computers.

As for the name "Monte Carlo," it is derived from that city in the Principality of Monaco famous for its ... casinos. The point is that one of the simplest mechanical devices for generating random numbers is the roulette wheel. We will discuss it in Chapter 2 under Generating Random Variables on a Computer. But it appears worthwhile to answer here one frequently asked question: "Does the Monte Carlo method help one win at roulette?" The answer is *No*; it is not even an attempt to do so.

Example: the "Hit-or-Miss" Method

We begin with a simple example. Suppose that we need to compute the area of a plane figure S. This may be a completely arbitrary figure with a curvilinear boundary; it may be defined graphically or analytically, and be either connected or consisting of several parts. Let S be the region drawn in Figure 1, and let us assume that it is contained completely within a unit square.

Choose at random N points in the square and designate the number of points that happen to fall inside S by N'. It is geometrically obvious that the area of S is approximately equal to the ratio N'/N. The greater the N, the greater the accuracy of this estimate.

The number of points selected in Figure 1 is $N = 40$. Of these, $N' = 12$ points appeared inside S. The ratio $N'/N = 12/40 = 0.30$, while the true area of S is 0.35.

In practice, the Monte Carlo method is not used for calculating the area of a plane figure. There are other methods (quadrature formulas) for this, that, though they are more complicated, provide much greater accuracy.

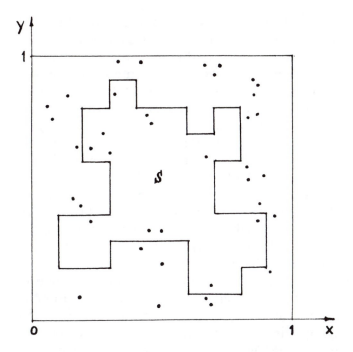

Fig. 1. N random points in the square. Of these, N' points are inside S. The area of S is approximately N'/N.

However, the hit-or-miss method shown in our example permits us to estimate, just as simply, the "multidimensional volume" of a body in a multidimensional space; in such a case the Monte Carlo method is often the only numerical method useful in solving the problem.

Two Distinctive Features
of the Monte Carlo Method

One advantageous feature of the Monte Carlo method is the simple structure of the computation algorithm. As a rule, a program is written to carry out one random trial (in our previous "hit-or-miss" example one has to check whether a selected random point inside the square also lies within S). This trial is repeated N times, each trial being independent of the rest, and then the results of all trials are averaged. Therefore, the Monte Carlo method is sometimes called the method of statistical trials.

A second feature of the method is that, as a rule, the error of calculations is proportional to $\sqrt{D/N}$, where D is some constant, and N is the number of trials. Hence, it is clear that to decrease the error by a factor of 10 (in other words, to obtain another decimal digit in the result), it is necessary to increase N (and thus the amount of work) by a factor of 100.

Obtaining high accuracy in this way is clearly impossible. Consequently, it is usually said that the Monte Carlo method is primarily useful for solving those problems that require moderate accuracy, e.g., 5 to 10%. However, any particular problem can be solved by different versions of the Monte Carlo method having different values of D. The accuracy of the result can be significantly improved by an ingenious choice of a computation method having a considerably smaller value of D.

The pluralized term "Monte Carlo methods" is frequently used, emphasizing that the same problem can be solved by simulating different random variables.

Problems that are Solvable by the Monte Carlo Method

To understand what kinds of problems are solvable by the Monte Carlo method, it is important to note that the method enables simulation of any process whose development is influenced by random factors. Second, for many mathematical problems involving no chance, the method enables us to artificially construct a probabilistic model (or several such models), making possible the solution of the problems. In fact, this was done in our earlier "hit-or-miss" example.

It is noteworthy that often, instead of simulating an actual random process, it is advantageous to use an artificial model. Such a situation is considered in Chapter 2 under Computation of Neutron Transmission Through a Plate and An Astrophysical Problem.

Thus, the Monte Carlo method is a universal numerical method for solving mathematical problems; its field of application is expanding with each new computer generation. The above-mentioned simple structure of Monte Carlo algorithms makes the method extremely convenient for multiprocessor (or parallel) computations.

More About the Example

Let us return to our hit-or-miss example, which requires the selection of random points in the unit square. How is this actually done?

Let us imagine the following experiment. An enlarged version of Figure 1 is hanging on a wall as a target. Some distance from the wall, a marksman shoots N times, aiming at the center of the square. We assume that the distance from the wall is sufficiently large (and the marksman is not the world champion). Then, of course, all the bullets will not hit the center; they will strike the target in N random points. Can we estimate the area of S from these points?

The result of such an experiment is shown in Figure 2. In this experiment $N = 40$, $N' = 24$ and the ratio $N'/N = 0.60$, which is almost double the actual area (0.35). It is obvious that if the marksman were very skilled, the result of the experiment would be very misleading because almost all of the bullets would hit the target near the center and thus inside S.

We can see that our method of computing area is valid when the random points are not only "random", but "uniformly distributed" over the whole square. To give these two terms precise meaning, we must be-

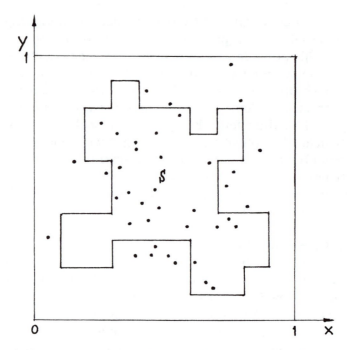

Fig. 2. N random hits in the square. Of these, N' hits inside S. Is the area approximately N'/N?

come acquainted with the definition of random variables and with some of their properties. This information is presented in the first part of Chapter 1 under Random Variables. A reader who is familiar with probability theory may omit this section, except for the discussion entitled The General Scheme of the Monte Carlo Method.

The procedures by which the random points in Figures 1 and 2 are actually computed are revealed at the end of Chapter 1 (see Again About the Hit-or-Miss Examples).

the author

Professor Ilya M. Sobol' was born in 1926 in Panevezys, Lithuania. He currently holds the position of Chief Researcher at the Institute for Mathematical Modeling, which is a division of the Russian Academy of Sciences, in Moscow.

His scientific interests include computational mathematics (primarily problems related to the Monte Carlo method), uniformly distributed sequences of points, multidimensional quadrature formulas, quasi-Monte Carlo methods, and multiple criteria decision-making.

Professor Sobol's publications include *The Method of Statistical Trials* (Monte Carlo method), Pergamon, 1964 (with N. P. Buslenko, D. I. Golenko, et al.); *Multidimensional Quadrature Formulas and Haar Functions*, Nauka, 1969 (Russian); *Numerical Monte Carlo Methods*, Nauka, 1973 (Russian); *Selection of Optimal Parameters in Problems with Several Criteria*, Nauka, 1981 (with R. B. Statnikov, Russian); *Comptonization and the Shaping of X-Ray Source Spectra: Monte Carlo Calculations*, Harwood Academic Publishers, 1983 (with R. A. Syunyaev and L. A. Pozdnyakov); *Points Which Uniformly Fill a Multidimensional Cube*, Znanie, 1985 (Russian); and *The Monte Carlo Method*, Nauka, 1968, 1972, 1978, 1985.

contents

A Primer for the Monte Carlo Method

simulating random variables

random variables

We assume that the reader is acquainted with the concept of probability, and we turn directly to the concept of a random variable.

The words "random variable," in ordinary lay usage, connote that one does not know what value a particular variable will assume. However, for mathematicians the term "random variable" has a precise meaning: though we do not know this variable's value in any given case, we do know the values it can assume and the probabilities of these values. The result of a single trial associated with this random variable cannot be precisely predicted from these data, but we can predict very reliably the result of a great number of trials. The more trials there are (the larger the sample), the more accurate our prediction.

Thus, to define a random variable, we must indicate the values it can assume and the probabilities of these values.

Discrete Random Variables

A random variable ξ is called *discrete* if it can assume any of a set of discrete values $x_1, x_2, \ldots, x_n$. A discrete random variable is therefore defined by a table

$$\xi \sim \begin{pmatrix} x_1 & x_2 & \cdots & x_n \\ p_1 & p_2 & \cdots & p_n \end{pmatrix} \qquad (T)$$

where $x_1, x_2, \ldots, x_n$ are the possible values of ξ, and $p_1, p_2, \ldots, p_n$ are the corresponding probabilities. To be precise, the probability that the random variable ξ will be equal to x_i (denoted by $\mathbf{P}\{\xi = x_i\}$) is equal to p_i:

$$\mathbf{P}\{\xi = x_i\} = p_i$$

Table (T) is called the *distribution* of the random variable ξ.

The values $x_1, x_2, \ldots, x_n$ can be arbitrary.* However, the probabilities $p_1, p_2, \ldots, p_n$ must satisfy two conditions:

1.   All p_i are positive:

$$p_i > 0 \qquad (1.1)$$

2.   The sum of all the p_i equals 1:

$$p_1 + p_2 + \ldots + p_n = 1 \qquad (1.2)$$

The latter condition requires that in each trial, ξ must necessarily assume one of the listed values.

The number

$$\mathbf{M}\xi = \sum_{i=1}^{n} x_i p_i \qquad (1.3)$$

is called the *mathematical expectation*, or the *expected value*, of the random variable ξ.

---

*In probability theory discrete random variables that can assume an infinite sequence of values are also considered.

To elucidate the physical meaning of this value, we rewrite it in the following form:

$$M\xi = \frac{\sum\limits_{i=1}^{n} x_i p_i}{\sum\limits_{i=1}^{n} p_i}$$

From this relation, we see that $M\xi$ is the average value of the variable ξ, in which more probable values are included with larger weights. (Averaging with weights is of course very common in science. In mechanics, for example, if masses m_1, m_2, ..., m_n are located at the points x_1, x_2, ..., x_n on the x axis, then the center of gravity of this system is given by the equation

$$\bar{x} = \frac{\sum\limits_{i=1}^{n} x_i m_i}{\sum\limits_{i=1}^{n} m_i}$$

Of course, in this case the sum of all the masses does not necessarily equal one.)

Let us mention the basic properties of the expected value. If c is an arbitrary nonrandom number, then

$$M(\xi + c) = M\xi + c \tag{1.4}$$

and

$$M(c\xi) = cM\xi \tag{1.5}$$

If ξ and η are two arbitrary random variables, then

$$M(\xi + \eta) = M\xi + M\eta \tag{1.6}$$

The number

$$D\xi = M([\xi - M\xi]^2) \tag{1.7}$$

is called the *variance* of the random variable ξ. Thus, the variance is the expected value of the squared deviation of the random variable ξ from its average value $M\xi$. Obviously, $D\xi$ is always greater than zero.

The expected value and the variance are the most important numerical characteristics of the random variable ξ. What is their practical value?

If we observe the variable ξ many times and obtain the values $\xi_1, \xi_2, \ldots, \xi_N$ (each of which is equal to one of the numbers $x_1, x_2, \ldots, x_n$), then the arithmetic mean of these values is close to $M\xi$:

$$\frac{1}{N}(\xi_1 + \xi_2 + \ldots + \xi_N) \approx M\xi \qquad (1.8)$$

and the variance $D\xi$ characterizes the spread of these values around the mean value $M\xi$.

Equation 1.8 is a simple case of the famous *law of large numbers* and can be explained by the following considerations. Assume that among the obtained values $\xi_1, \xi_2, \ldots, \xi_N$ the number x_1 occurs k_1 times, the number x_2 occurs k_2 times, $\ldots$, and the number x_n occurs k_n times. Then

$$\sum_{j=1}^{N} \xi_j = x_1 k_1 + x_2 k_2 + \ldots + x_n k_n$$

Hence,

$$\frac{1}{N}\sum_{j=1}^{N} \xi_j = x_1\frac{k_1}{N} + x_2\frac{k_2}{N} + \cdots + x_n\frac{k_n}{N}$$

At large N, the frequency k_i/N of the value x_i approaches its probability p_i so that $k_i/N \approx p_i$. Therefore,

$$\frac{1}{N}\sum_{j=1}^{N} \xi_j = \sum_{i=1}^{n} x_i\frac{k_i}{N} \approx \sum_{i=1}^{n} x_i p_i = M\xi$$

Equation 1.7 for the variance can be transformed using Equations 1.4, 1.5, and 1.6:

$$D\xi = M(\xi^2 - 2\xi \cdot M\xi + [M\xi]^2)$$

$$= M(\xi^2) - 2M\xi \cdot M\xi + [M\xi]^2$$

from which it follows that

$$D\xi = M(\xi^2) - (M\xi)^2 \qquad (1.9)$$

Usually the computation of variance by Equation 1.9 is simpler than by Equation 1.7.

The variance has the following basic properties. If c is an arbitrary nonrandom number, then

$$D(\xi + c) = D\xi \qquad (1.10)$$

and

$$D(c\xi) = c^2 D\xi \qquad (1.11)$$

The concept of *independence* of random variables plays an important role in probability theory. Independence is a rather complicated concept, though it may be quite clear in the simplest cases. Let us suppose that we are simultaneously observing two random variables ξ and η. If the distribution of ξ does not change when we know the value that η assumes, then it is natural to consider ξ independent of η.

The following relations hold for independent random variables ξ and η:

$$M(\xi\eta) = M\xi \cdot M\eta \qquad (1.12)$$

and

$$D(\xi + \eta) = D\xi + D\eta \qquad (1.13)$$

Example: Throwing a Die

Let us consider a random variable $\varkappa$ with distribution specified by the table

$$\varkappa \sim \begin{pmatrix} 1 & 2 & 3 & 4 & 5 & 6 \\ 1/6 & 1/6 & 1/6 & 1/6 & 1/6 & 1/6 \end{pmatrix}$$

Clearly, $\varkappa$ can assume the values 1, 2, 3, 4, 5, 6, and each of these values is equally probable. So, the number of pips appearing when a die is thrown can be used to calculate $\varkappa$.

According to Equation 1.3

$$\mathbf{M}x = 1 \cdot \frac{1}{6} + 2 \cdot \frac{1}{6} + \cdots + 6 \cdot \frac{1}{6} = 3.5$$

and according to Equation 1.9

$$\mathbf{D}x = \mathbf{M}(x^2) - (\mathbf{M}x)^2 = 1^2 \cdot \frac{1}{6} + 2^2 \cdot \frac{1}{6} + \cdots$$

$$+ 6^2 \cdot \frac{1}{6} - (3.5)^2 = 2.917$$

Example: Tossing a Coin

Let us consider a random variable θ with distribution

$$\theta \sim \begin{pmatrix} 3 & 4 \\ 1/2 & 1/2 \end{pmatrix}$$

The game of tossing a coin, with the agreement that a head counts three points and a tail counts four points, can be used to generate θ. Here

$$\mathbf{M}\theta = 3 \cdot \frac{1}{2} + 4 \cdot \frac{1}{2} = 3.5$$

and

$$\mathbf{D}\theta = 3^2 \cdot \frac{1}{2} + 4^2 \cdot \frac{1}{2} - (3.5)^2 = 0.25$$

We see that $\mathbf{M}\theta = \mathbf{M}x$, but $\mathbf{D}\theta < \mathbf{D}x$. This could easily have been predicted, since the maximum deviation of θ from 3.5 is ±0.5, while for the values of x, the spread can reach ±2.5.

Continuous Random Variables

Let us assume that some radium is placed on a Cartesian plane at the origin. As an atom of radium decays, an α-particle is emitted. Its direction is described by the angle ψ (Figure 1.1). Since, both in theory and practice, any direction of emission is possible, this random variable can assume any value from 0 to 2π.

We shall call a random variable *continuous* if it can assume any value in a certain interval (a, b).

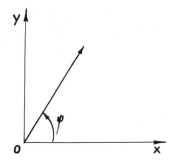

Fig. 1.1. Random direction.

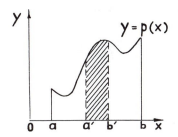

Fig. 1.2. Probability density.

A continuous random variable ξ is defined by specifying an interval containing all its possible values, and a function $p(x)$ that is called the *probability density* of the random variable ξ (or the *distribution density* of ξ).

The physical meaning of $p(x)$ is as follows: let (a', b') be an arbitrary interval contained in (a, b) (that is, $a \leq a'$, $b' \leq b$). Then the probability that ξ falls in the interval (a', b') is equal to the integral

$$\mathbf{P}\{a' < \xi < b'\} = \int_{a'}^{b'} p(x)\, dx \qquad (1.14)$$

This integral (1.14) is equal to the shaded area in Figure 1.2.

The set of values of ξ can be any interval. The cases $a = -\infty$ and/or $b = \infty$ are also possible. However, the density $p(x)$ must satisfy two conditions analogous to conditions (1.1) and (1.2) for discrete variables:

1.   The density $p(x)$ is positive inside (a, b):

$$p(x) > 0 \qquad\qquad (1.15)$$

2.   The integral of the density $p(x)$ over the whole interval (a, b) is equal to 1:

$$\int_a^b p(x)\,dx = 1 \qquad\qquad (1.16)$$

The number

$$M\xi = \int_a^b xp(x)\,dx \qquad\qquad (1.17)$$

is called the *expected value* of the continuous random variable ξ.

The expected value has the same meaning as in the case of a discrete random variable. Indeed, since

$$M\xi = \frac{\int_a^b xp(x)\,dx}{\int_a^b p(x)\,dx}$$

it can easily be seen that this is the average value of ξ: any value of x from the interval (a, b) enters the integral with its weight $p(x)\,dx$.

(In this case we also have an analogous equation in mechanics: if the linear density of a rod $a < x < b$

is equal to $\rho(x)$, then the abscissa of the center of gravity is given by

$$\bar{x} = \frac{\int\limits_a^b x\rho(x)\,dx}{\int\limits_a^b \rho(x)\,dx}$$

Of course, the total mass of the rod $\int\limits_a^b \rho(x)\,dx$ does not necessarily equal one.)

All the Equations 1.4 through 1.13 are valid also for continuous random variables. This includes the definition of variance 1.7, the Equation 1.9, and all the properties of $M\xi$ and $D\xi$.*

We will mention just one more formula for the expectation of a function of ξ. Again, let the random variable ξ have probability density $p(x)$. Consider an arbitrary continuous function $f(x)$ and define a random variable $\eta = f(\xi)$. It can be proved that

$$Mf(\xi) = \int\limits_a^b f(x)p(x)\,dx \qquad (1.18)$$

(Of course, a similar equation is valid for a discrete random variable ξ with distribution (T): $Mf(\xi) = \sum\limits_{i=1}^n f(x_i)p_i$.)

It must be stressed that, in general, $Mf(\xi) \neq f(M\xi)$.

Finally, for a continuous random variable ξ and an arbitrary value x

$$P\{\xi = x\} = 0$$

Therefore, the probability of an equality $\{\xi = x\}$ is physically meaningless. Physically meaningful are

---

*However, in probability theory more general random variables are encountered: where the condition (1.15) is weakened to $p(x) \geq 0$, where the expected value $M\xi$ does not exist, and where the variance $D\xi$ is infinite.

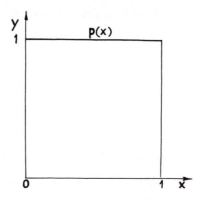

Fig. 1.3. Constant density
(uniform distribution).

probabilities of falling into a small interval:

$$\mathbf{P}\{x \leq \xi < x + dx\} = p(x)\,dx$$

An Important Example

A random variable γ defined over the interval $0 < x < 1$ and having a density $p(x) = 1$ is said to be *uniformly distributed* over $(0, 1)$ (Figure 1.3). For any subinterval (a', b') within $(0, 1)$, the probability that γ lies in (a', b') is equal to

$$\int_{a'}^{b'} p(x)\,dx = b' - a'$$

that is, the length of the subinterval. In particular, if we divide $(0, 1)$ into any number of intervals of equal length, the probabilities of γ falling into any of these intervals are the same.

It is easy to calculate that

$$\mathbf{M}\gamma = \int_{0}^{1} xp(x)\,dx = \int_{0}^{1} x\,dx = 1/2$$

and

$$\mathbf{D}\gamma = \int_0^1 x^2 p(x)\, dx - (\mathbf{M}\gamma)^2 = 1/3 - 1/4 = 1/12$$

This random variable γ will be used frequently below.

Normal Random Variables

A *normal* (or *Gaussian*) random variable is a random variable ζ defined on the whole axis $-\infty < x < \infty$ and having the density

$$p(x) = \frac{1}{\sigma\sqrt{2\pi}} \exp\left(-\frac{(x-a)^2}{2\sigma^2}\right) \qquad (1.19)$$

where a and $\sigma > 0$ are real parameters. (The character "σ" in this equation represents a number and not a random variable; the use of a Greek letter here is traditional. Equation 1.19 can be found on the German ten-mark banknote beside the portrait of C. F. Gauss.)

The parameter a does not affect the shape of the curve $p(x)$: varying a results only in displacement of the curve along the x axis. However, a change in σ does change the shape of the curve. Indeed, it is easy to see that

$$\max p(x) = p(a) = \frac{1}{\sigma\sqrt{2\pi}}$$

If σ decreases, the $\max p(x)$ increases. But, according to condition (1.16), the entire area under the curve $p(x)$ is equal to 1. Therefore, the curve extends upward near $x = a$, but decreases for all sufficiently large values of x. Two normal densities corresponding to $a = 0$, with $\sigma = 1$ and $a = 0$, with $\sigma = 0.5$ are drawn in Figure 1.4. (Another normal density can be found in Figure 2.6 below.)

It can be proved that

$$\mathbf{M}\zeta = a, \text{ and } \mathbf{D}\zeta = \sigma^2$$

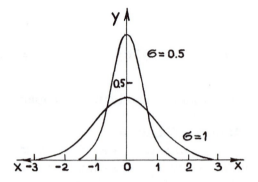

Fig. 1.4. Two Gaussian (or normal) densities.

Arbitrary probabilities $\mathbf{P}\{x' < \zeta < x''\}$ can be easily computed using one table containing values of the function

$$\Phi(x) = \frac{2}{\sqrt{2\pi}} \int\limits_0^x e^{-t^2/2}\, dt$$

that is usually called the *error function*. This term is sometimes used for other functions (e.g., erf x), but all such functions can easily be transformed into $\Phi(x)$. To show this, in accordance with (1.14), we write

$$\mathbf{P}\{x' < \zeta < x''\} = \frac{1}{\sigma\sqrt{2\pi}} \int\limits_{x'}^{x''} \exp\left(-\frac{(x-a)^2}{2\sigma^2}\right)\, dx$$

In the integral we make the substitution $x - a = \sigma t$. This produces

$$\mathbf{P}\{x' < \zeta < x''\} = \frac{1}{\sqrt{2\pi}} \int\limits_{t_1}^{t_2} \exp\left(-t^2/2\right)\, dt$$

where $t_1 = (x' - a)/\sigma$ and $t_2 = (x'' - a)/\sigma$. Hence,

$$\mathbf{P}\{x' < \zeta < x''\} = \frac{1}{2}\left(\Phi(t_2) - \Phi(t_1)\right)$$

Tables of $\Phi(x)$ contain only positive values of x since $\Phi(-x) = -\Phi(x)$.

Normal random variables are encountered in a wide variety of problems. (The reason for this will be explained later in this chapter in a discussion of the central limit theorem.) Here we shall consider two different approaches to estimating the deviations of ζ from $a = M\zeta$.

the rule of "three sigmas"

Assume that $x' = a - 3\sigma$, $x'' = a + 3\sigma$. Then $t_1 = -3$, $t_2 = 3$, and from the last formula it follows that

$$\mathbf{P}\{a - 3\sigma < \zeta < a + 3\sigma\} = \Phi(3) = 0.997 \qquad (1.20)$$

The probability 0.997 is so near to 1 that often Equation 1.20 is given the following interpretation: for a single trial it is practically impossible to obtain a value of ζ differing from $M\zeta$ by more than 3σ.

the probable error

Consider now $x' = a - r$, $x'' = a + r$ where the quantity r is defined as

$$r = 0.6745\sigma$$

Then $t_1 = -0.6745$, $t_2 = 0.6745$ and

$$\mathbf{P}\{a - r < \zeta < a + r\} = \Phi(0.6745) = 0.500$$

This relation can be rewritten in the form

$$\mathbf{P}\{|\zeta - a| < r\} = 0.5$$

Hence, the probability of the opposite inequality is also 0.5:

$$\mathbf{P}\{|\zeta - a| > r\} = 0.5$$

(of course, $\mathbf{P}\{|\zeta - a| = r\} = 0$ since ζ is a continuous random variable).

The last two relations show that values of ζ deviating from a by more than r and values deviating from a by less than r are equally probable. Therefore, r is called the *probable error* of ζ.

Example: Error of Measurement

The error δ in an experimental measurement is usually a normal random variable. If there is no systematic error (bias) then $a = M\delta = 0$. But the quantity $\sigma = \sqrt{D\delta}$, called the *standard deviation* of δ, is always positive, and characterizes the error of the method of measurement (the precision of the method).

For a single measurement the absolute error $|\delta|$ as a rule does not exceed 3σ. The probable error $r = 0.6745\sigma$ shows the order of magnitude of $|\delta|$ that can be both smaller or larger than r:

$$\mathbf{P}\{|\delta| \leq r\} = \mathbf{P}\{|\delta| \geq r\} = 0.5$$

The Central Limit Theorem of Probability Theory

This remarkable theorem was first formulated by P. S. Laplace. Many outstanding mathematicians, including the Russians P. L. Chebyshev, A. A. Markov, A. M. Lyapunov, and A. Ya. Khinchin, have worked on various generalizations of the original theorem. All the proofs are rather complex.

Let us consider N identical independent random variables $\xi_1, \xi_2, \ldots, \xi_N$, so that their probability distributions coincide. Consequently, their mathematical expectations and variances also coincide (we assume that they are finite). The random variables can be continuous or discrete.

Let us designate

$$M\xi_1 = M\xi_2 = \ldots = M\xi_N = m$$

$$D\xi_1 = D\xi_2 = \ldots = D\xi_N = b^2$$

Denote the sum of all these variables by ρ_N:

$$\rho_N = \xi_1 + \xi_2 + \ldots + \xi_N$$

From Equations 1.6 and 1.13 it follows that

$$M\rho_N = M(\xi_1 + \xi_2 + \ldots + \xi_N) = Nm$$

and
$$\mathbf{D}\rho_N = \mathbf{D}(\xi_1 + \xi_2 + \ldots + \xi_N) = Nb^2$$

Now let us consider a normal random variable ζ_N with the same parameters: $a = Nm$, $\sigma = b\sqrt{N}$. Its density is denoted $p_N(x)$. The central limit theorem states that for any interval (x', x'') and for all large N

$$\mathbf{P}\{x' < \rho_N < x''\} \approx \int_{x'}^{x''} p_N(x)\, dx$$

The physical meaning of this theorem is clear: *the sum ρ_N of a large number of identical independent random variables is approximately normal.* Actually, this theorem is valid under much weaker conditions: the variables $\xi_1, \xi_2, \ldots, \xi_N$ should not necessarily be identical and independent; essentially, all that is required is that individual variables ξ_j do not play too great a role in the sum.

It is this theorem that explains why normal random variables are so often encountered in nature. Indeed, whenever we meet an aggregate effect of a large number of small random factors, the resulting random variable is normal. For example, the scattering of artillery shells is almost always normal, since it depends on weather conditions in all the various regions of the trajectory as well as on many other factors.

The General Scheme
of the Monte Carlo Method

Suppose that we need to calculate some unknown quantity m. Let us try to find a random variable ξ with $\mathbf{M}\xi = m$. Assume that the variance of ξ is $\mathbf{D}\xi = b^2$.

Consider N independent random variables $\xi_1, \xi_2, \ldots, \xi_N$ with distributions identical to that of ξ. If N is sufficiently large, then it follows from the central limit theorem that the distribution of the sum

$$\rho_N = \xi_1 + \xi_2 + \ldots + \xi_N$$

will be approximately normal, with $a = Nm$ and $\sigma = b\sqrt{N}$. According to the rule of "three sigmas" (1.20)

$$\mathbf{P}\{Nm - 3b\sqrt{N} < \rho_N < Nm + 3b\sqrt{N}\} \approx 0.997$$

If we divide the inequality within the parentheses by N, we obtain an equivalent inequality, whose probability remains the same:

$$\mathbf{P}\{m - \frac{3b}{\sqrt{N}} < \frac{\rho_N}{N} < m + \frac{3b}{\sqrt{N}}\} \approx 0.997$$

We can rewrite the last expression in a somewhat different form:

$$\mathbf{P}\left\{\left|\frac{1}{N}\sum_{j=1}^{N}\xi_j - m\right| < \frac{3b}{\sqrt{N}}\right\} \approx 0.997 \qquad (1.21)$$

This is an extremely important relation for the Monte Carlo method, giving us both the method for calculating m and the error estimate.

Indeed, we have to find N values of the random variable ξ — selecting one value of each of the variables $\xi_1, \xi_2, \ldots, \xi_N$ is equivalent to selecting N values of ξ, since all these variables have identical distributions.

From (1.21) it is obvious that the arithmetic mean of these values will be approximately equal to m. In all likelihood, the error of this approximation does not exceed $3b/\sqrt{N}$, and approaches zero as N increases.

In practical computations, the error bound $3b/\sqrt{N}$ is often loose, and it is more convenient to use the probable error

$$r_N = 0.6745b/\sqrt{N}$$

However, this is not a bound — this is a characteristic of the absolute error

$$\left|\frac{1}{N}\sum_{j=1}^{N}\xi_j - m\right|$$

generating random variables on a computer

Sometimes the problem statement of generating random variables on a computer provokes the question: "Everything the machine does must be programmed beforehand, so where can randomness come from?" There are, indeed, certain difficulties associated with this point, but they belong more to philosophy than to mathematics, and we will not consider them here.

We will only stress that random variables are ideal mathematical concepts. Whether natural phenomena can actually be described by means of these variables can only be ascertained experimentally. Such a description is always approximate. Moreover, a random variable that satisfactorily describes a physical quantity in one type of phenomenon may prove unsatisfactory when used to describe the same quantity in other phenomena. Analogously on a national map a road may be depicted as a straight line, whereas on a local city map the same road must be drawn as a twisted band.

Usually, three means for obtaining random variables are considered: tables of random numbers, random number generators, and the pseudorandom number method. We will discuss each.

Tables of Random Numbers

Let us perform the following experiment. We mark the digits $0, 1, 2, \ldots, 9$ on ten identical slips of paper, place them in a hat, mix them, and take one out; then return it and mix again. We write down the digits obtained in this way in a table like Table 1.1 (the digits in Table 1.1 are arranged in groups of five for convenience).

Such a table is usually called a *table of random numbers*, though *random digits* would be a better

Table 1.1. 400 Random Digits

86515	90795	66155	66434	56558	12332	94377	57802
69186	03393	42502	99224	88955	53758	91641	18867
41686	42163	85181	38967	33181	72664	53807	00607
86522	47171	88059	89342	67248	09082	12311	90316
72587	93000	89688	78416	27589	99528	14480	50961
52452	42499	33346	83935	79130	90410	45420	77757
76773	97526	27256	66447	25731	37525	16287	66181
04825	82134	80317	75120	45904	75601	70492	10274
87113	84778	45863	24520	19976	04925	07824	76044
84754	57616	38132	64294	15218	49286	89571	42903

term. This table can be put into a computer's memory. Then, when performing a calculation, if we require values of a random variable ε with the distribution

$$\varepsilon \sim \begin{pmatrix} 0 & 1 & 2 & \ldots & 9 \\ 0.1 & 0.1 & 0.1 & \ldots & 0.1 \end{pmatrix} \qquad (1.22)$$

then we need only take the next digit from this table.

The largest of all published tables of random numbers contains one million random digits (see RAND Corporation).[2] Of course, it was compiled with the aid of technical equipment more sophisticated than a hat: a special electronic roulette wheel was constructed. Figure 1.5 shows an elementary scheme of such a roulette wheel.

It should be noted that a good table of random numbers is not as easy to compile as it may initially appear. Any real physical device produces random numbers with distributions that differ slightly from ideal distributions (1.22); in addition, experimental errors may occur (for example, a slip of paper might stick to the hat's lining). Therefore, compiled tables are carefully examined, using special statistical tests, to check whether any properties of the group of num-

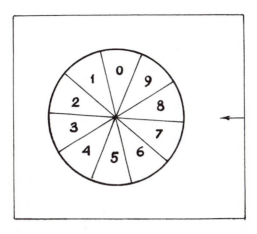

Fig. 1.5. Roulette wheel for generating random digits (scheme).

bers contradict the hypothesis that these numbers are independent values of the random variable ε.

Let us examine the simplest, and, at the same time, most important tests. Consider a table containing N digits $\varepsilon_1, \varepsilon_2, \ldots, \varepsilon_N$. Denote the number of zeros in this table by ν_0, the number of ones by ν_1, the number of twos by ν_2, and so on. Consider the sum

$$\sum_{i=0}^{9} \left(\nu_i - \frac{N}{10} \right)^2$$

Probability theory enables us to predict the range in which this sum should lie; its value should not be too large, since the expected value of each of the ν_i is equal to $N/10$, but neither should it be too small, since that would indicate a "too regular" distribution of the numbers. ("Too regularly" distributed numbers facilitate certain computations, known as *quasi-Monte Carlo methods*. But these numbers cannot be used as general purpose random numbers.)

Assume now that the number N is even, $N = 2N'$, and consider pairs $(\varepsilon_1, \varepsilon_2), (\varepsilon_3, \varepsilon_4), \ldots, (\varepsilon_{N-1}, \varepsilon_N)$. Denote by ν_{ij} the number of pairs equal to (i, j) and calculate the sum

$$\sum_{i=0}^{9}\sum_{j=0}^{9}\left(\nu_{ij} - \frac{N'}{100}\right)^2$$

Again, probability theory predicts the range in which this sum should lie, and thus we can test the distribution of pairs. (Similarly, we may test the distribution of triplets, quadruplets, etc.)

However, tables of random numbers are used only for Monte Carlo calculations performed by hand. Computers cannot store such large tables in their small internal memories, and storing such tables in a computer's external memory considerably slows calculations.

Generators of Random Numbers

It would seem that a roulette wheel could be coupled to a computer, in order to generate random numbers as needed. However, because any such mechanical device would be too slow for a computer, vacuum tube noise is usually proposed as a source of randomness. For example, in Figure 1.6, the noise level E is monitored; if, within some fixed time interval Δt, the noise exceeds a given threshold E_0 an even number of times, then a zero is recorded; if the noise exceeds E_0 an odd number of times, a one is recorded. (More sophisticated devices also exist.)

At first glance this appears to be a very convenient method. Let m such generators work in parallel, all the time, and send random zeros and ones into a particular address in RAM. At any moment the computer can refer to this cell and take from it a value of the random variable γ distributed uniformly over the interval $0 < x < 1$. This value is, of course, approximate, being an m-digit binary fraction of the form $0.\alpha_1\alpha_2\ldots\alpha_m$, where each α_i simulates a random vari-

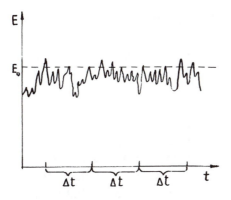

Fig. 1.6. Random noise for generating random bits (scheme).

able with the distribution

$$\alpha \sim \begin{pmatrix} 0 & 1 \\ 1/2 & 1/2 \end{pmatrix}$$

But this method is not free of defects. First, it is diffi-cult to check the "quality" of the numbers produced. Tests must be carried out periodically, since any im-perfection can lead to a "distribution drift" (that is, zeros and ones in some places begin to appear with unequal frequencies). Second, it is desirable to be able to repeat a calculation on the computer, but im-possible to reproduce the same random numbers if they are not stored throughout the calculation; as discussed earlier, storing so much data is impracti-cal.

Random number generators may prove useful if specialized computers are ever designed for solving problems by means of the Monte Carlo method. But it is simply not economical to install and maintain such a special unit in multipurpose computers, in which computations involving random numbers are performed only occasionally. It is therefore better to use pseudorandom numbers.

Pseudorandom Numbers

Since the "quality" of random numbers used for computations is checked by special tests, one can ignore the means by which random numbers are produced, as long as they satisfy the tests. We may even try to calculate random numbers by a prescribed — albeit sophisticated — formula.

Numbers obtained by a formula that simulate the values of the random variable γ are called *pseudorandom numbers*. The word "simulate" means that these numbers satisfy a set of tests just as if they were independent values of γ.

The first algorithm for generating pseudorandom numbers, the *mid-square method*, was proposed by John von Neumann. We illustrate it with an example.

Suppose we are given a four-digit number $\gamma_0 = 0.9876$. We square it and obtain an eight-digit number $\gamma_0^2 = 0.97535376$. Then we take out the middle four digits of this number and get $\gamma_1 = 0.5353$.

Now we square γ_1 and obtain $\gamma_1^2 = 0.28654609$. Once more we take out the middle four digits, and get $\gamma_2 = 0.6546$.

Then we obtain $\gamma_2^2 = 0.42850116$, $\gamma_3 = 0.8501$; $\gamma_3^2 = 0.72267001$, $\gamma_4 = 0.2670$; $\gamma_4^2 = 0.07128900$, $\gamma_5 = 0.1289$; and so on.

Unfortunately, this algorithm tends to produce a disproportionate frequency of small numbers; however, other, better, algorithms have been discovered — these will be discussed in Chapter 3 under On Pseudorandom Numbers.

The advantages of the pseudorandom numbers method are evident. First, obtaining each number requires only a few simple operations, so the speed of generating numbers is of the same order as the computer's work speed. Second, the program occupies only a few addresses in RAM. Third, any of the numbers γ_k can be reproduced easily. And finally, the "quality" of this sequence need be checked only once;

after that, it can be used many times in calculations of similar problems without taking any risk.

The single shortcoming of the method is the limited supply of pseudorandom numbers that it gives, since if the sequence of numbers $\gamma_0, \gamma_1, \ldots, \gamma_k, \ldots$ is computed by an algorithm of the form

$$\gamma_k = \mathbf{F}(\gamma_{k-1})$$

it must be periodic. (Indeed, in any address in RAM only a finite number of different numbers can be written. Thus, sooner or later one of the numbers, say γ_L, will coincide with one of the preceding numbers, say γ_l. Then clearly, $\gamma_{L+1} = \gamma_{l+1}, \gamma_{L+2} = \gamma_{l+2}, \ldots$, so that there is a period $P = L - l$. The nonperiodic part of the sequence is $\gamma_1, \gamma_2, \ldots, \gamma_{L-1}$.)

A large majority of computations performed by the Monte Carlo method use sequences of pseudorandom numbers whose periods exceed current requirements.

A Remark

One must be careful: there are papers, books, and software advertizing inadequate methods for generating pseudorandom numbers, these numbers being checked only with the simplest tests. Attempts to use such numbers for solving more complex problems lead to biased results (see On Pseudo Random Numbers in Chapter 3).

transformations of random variables

In the early stage of application of the Monte Carlo method, some users tried to construct a special roulette for each random variable. For example, a "roulette wheel" disk divided into unequal parts (proportional to p_i), shown in Figure 1.7, can be used to generate values of a random variable with the distri-

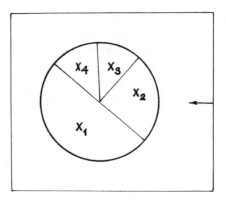

Fig. 1.7. Roulette wheel for generating values of a discrete random variable.

bution

$$\begin{pmatrix} x_1 & x_2 & x_3 & x_4 \\ 0.5 & 0.25 & 0.125 & 0.125 \end{pmatrix}$$

However, this proves unnecessary: values of any random variable can be obtained by transformations of values of one "standard" random variable. Usually this role is played by γ, which is uniformly distributed over the interval $0 < x < 1$. We already know how to get the values of γ; the process of finding a value of some random variable ξ, accomplished by transforming one or more values of γ, we call the *modeling* of ξ.

Modeling of a Discrete Random Variable

Assume that we want to obtain values of a random variable ξ with the distribution

$$\xi \sim \begin{pmatrix} x_1 & x_2 & \dots & x_n \\ p_1 & p_2 & \dots & p_n \end{pmatrix}$$

Consider the interval $0 < y < 1$, and break it up into n intervals with lengths equal to $p_1, p_2, \dots, p_n$. The coordinates of the points of division will obviously be $y = p_1, y = p_1 + p_2, y = p_1 + p_2 + p_3, \dots, y = p_1 + p_2 +$

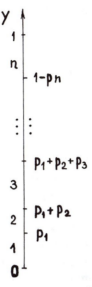

Fig. 1.8. Partition of the interval $0 < y < 1$ for modeling of a discrete random variable.

$\ldots + p_{n-1}$. These intervals are numbered $1, 2, \ldots, n$ (Figure 1.8), and this completes the preparation for modeling ξ. Each time we "perform an experiment" in order to find a value of ξ, we select a value of γ and fix the point $y = \gamma$. If this point falls into the interval numbered i, we consider that $\xi = x_i$.

It is easy to demonstrate the validity of this procedure. Since γ is uniformly distributed in the unit interval, the probability of γ falling into any subinterval is equal to the length of that subinterval. Hence,

$$\mathbf{P}\{0 < \gamma < p_1\} = p_1$$

$$\mathbf{P}\{p_1 < \gamma < p_1 + p_2\} = p_2$$

$$\ldots$$

$$\mathbf{P}\{p_1 + \ldots + p_{n-1} < \gamma < 1\} = p_n$$

According to our procedure, $\xi = x_i$ if

$$p_1 + \ldots + p_{i-1} < \gamma < p_1 + \ldots + p_i$$

and the probability of this event is p_i.

In practice, the definition of i can be carried out with the aid of comparisons: is γ less than p_1? Otherwise, is γ less than $p_1 + p_2$? Otherwise, is γ less than $p_1 + p_2 + p_3$? And so on.

Note that the order of the values x_1, x_2, $\ldots$, x_n in the distribution of ξ can be arbitrary, but it must be fixed before the modeling.

equiprobable values

The previous modeling procedure is much simpler when all the probabilities p_i are equal: $p_1 = p_2 = \ldots = p_n = 1/n$; that is, when

$$\xi \sim \begin{pmatrix} x_1 & x_2 & \ldots & x_n \\ 1/n & 1/n & \ldots & 1/n \end{pmatrix}$$

According to our procedure, $\xi = x_i$ if $(i-1)/n < \gamma < i/n$ or, in other words, $i - 1 < n\gamma < i$. The last relation means that the integral part of $n\gamma$ is equal to $i - 1$.

The integral part of z is usually denoted $[z]$. So, we obtain a direct formula for modeling ξ:

$$\xi = x_i \text{ where } \quad i = [n\gamma] + 1$$

numerical example

Find 10 values of the random variable $\varkappa$ considered in the discussion of throwing a die, whose possible values are $x_i = i$, $1 \leq i \leq 6$. As values of γ, we select groups of three random digits, from Table 1, multiplied by 0.001. So, $\gamma = 0.865$; 0.159; 0.079; 0.566; 0.155; 0.664; 0.345; 0.655; 0.812; 0.332. The corresponding values $\varkappa = 1 + [6\gamma]$ are 6; 1; 1; 4; 1; 4; 3; 4; 5; 2. They may be regarded as ten throws of a die.

Modeling of a Continuous Random Variable

Now let us assume that we want to obtain values of a random variable ξ distributed over the interval $a < x < b$ with density $p(x)$. We shall prove that values of ξ can be found from the equation

$$\int_a^\xi p(x)\, dx = \gamma \qquad (1.23)$$

That is, selecting a consecutive value of γ, we must solve Equation 1.23 to find the corresponding value of ξ.

For the proof let us consider the function

$$y = \int_a^x p(x)\, dx$$

From the general properties of density (1.15) and (1.16), it follows that

$$y(a) = 0, \quad y(b) = 1$$

and the derivative

$$y'(x) = p(x) > 0$$

This means that the function $y(x)$ increases monotonically from 0 to 1 (Figure 1.9). Any straight line $y = \gamma$, where $0 < \gamma < 1$, intersects the curve $y = y(x)$ in one, and only one, point, whose abscissa is taken for the value of ξ. Thus Equation 1.23 always has a unique solution.

Now we select an arbitrary interval (a', b') contained in (a, b). The ordinates of the curve $y = y(x)$ satisfying the inequality

$$y(a') < y < y(b')$$

correspond to the points of this interval

$$a' < x < b'$$

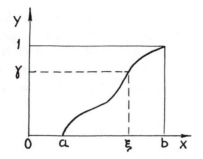

Fig. 1.9. The function $y = \int\limits_{a}^{x} p(x)\,dx$ for modeling of a continuous random variable.

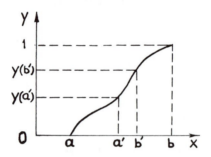

Fig. 1.10. One-to-one mapping of the interval $a' < x < b'$ onto $y(a') < y < y(b')$.

Consequently, if ξ belongs to $a' < x < b'$, then γ belongs to $y(a') < y < y(b')$, and vice versa (Figure 1.10). Hence,

$$\mathbf{P}\{a' < \xi < b'\} = \mathbf{P}\{y(a') < \gamma < y(b')\}$$

Since γ is uniformly distributed over $0 < y < 1$,

$$\mathbf{P}\{y(a') < \gamma < y(b')\} = y(b') - y(a') = \int_{a'}^{b'} p(x)\,dx$$

Therefore,

$$\mathbf{P}\{a' < \xi < b'\} = \int_{a'}^{b'} p(x)\,dx$$

and this means that the random variable ξ, which is a root of Equation 1.23, has the probability density $p(x)$.

Example

The random variable η is said to be uniformly distributed over the interval (a, b) if its density is constant over this interval:

$$p(x) = \frac{1}{b-a} \quad \text{for all } a < x < b$$

To model values of η, we apply Equation 1.23:

$$\int_{a}^{\eta} \frac{dx}{b-a} = \gamma$$

The integral is easily computed:

$$\frac{\eta - a}{b - a} = \gamma$$

Hence, we arrive at an explicit expression

$$\eta = a + \gamma(b - a) \tag{1.24}$$

that is a linear transformation of γ.

Other examples of the application of Equation 1.23 are given in Chapter 2 (see The Simple Flow of Requests and A Numerical Example).

von Neumann's Method for Modeling a Continuous Random Variable

It may happen that Equation 1.23 is difficult to solve for ξ (for example, when the integral of $p(x)$ cannot be expressed in terms of elementary functions, or when the density $p(x)$ is given graphically).

Let us assume that the random variable ξ is defined in a finite interval $a < x < b$, and its density is bounded (Figure 1.11):

$$p(x) \leq M_0$$

The value of ξ can be found in the following way:

1. We select two random numbers γ' and γ'' and construct a random point $\Gamma(\eta', \eta'')$ with coordinates

$$\eta' = a + \gamma'(b - a), \quad \eta'' = \gamma'' M_0$$

2. If the point Γ lies under the curve $y = p(x)$, we set $\xi = \eta'$; if the point Γ lies above the curve, we reject the pair (γ', γ'') and select a new pair (γ', γ'').

The validity of this method can be demonstrated easily. First, since η' is uniformly distributed in (a, b), the probability that Γ will be in the strip $(x, x + dx)$ is proportional to dx. Second, since η'' is uniformly distributed in $(0, M_0)$, the probability that Γ will not be rejected is equal $p(x)/M_0$ and so proportional to $p(x)$. Hence, the probability that an accepted value $\xi = \eta'$ belongs to the interval $(x, x+dx)$ is proportional to $p(x)\,dx$.

On Modeling Normal Variables

There are alternative techniques for modeling various random variables. These techniques are usually applied only when the methods for modeling discrete and continuous random variables described earlier

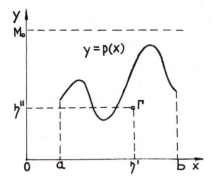

Fig. 1.11. von Neumann's method: $\xi = \eta'$.

prove ineffective (see On Methods for Generating Random Variables in Chapter 3).

In particular, these alternative techniques are applied in the case of a normal random variable ζ, since the equation

$$\frac{1}{\sqrt{2\pi}} \int\limits_{-\infty}^{\zeta} \exp\left(-\frac{x^2}{2}\right) dx = \gamma$$

is not explicitly solvable.

For computations by hand, one can use Table 1.2, in which values are given for a normal random variable ζ with mathematical expectation $M\zeta = 0$ and variance $D\zeta = 1$. It is not hard to verify that the random variable

$$\zeta' = a + \sigma\zeta \tag{1.25}$$

will also be normal; as follows from Equations 1.4, 1.5, 1.10, and 1.11

$$M\zeta' = a, \quad D\zeta' = \sigma^2$$

Thus, Equation 1.25 enables us to model arbitrary normal variables with the help of Table 1.2.

Table 1.2. 88 Normal Values with $a = 0$, $\sigma = 1$

0.2005	1.1922	−0.0077	0.0348	1.0423	−1.8149	1.1803
0.0033	1.1609	−0.6690	−1.5893	0.5816	1.8818	0.7390
−0.2736	1.0828	0.5864	−0.9245	0.0904	1.5068	−1.1147
0.2776	0.1012	−1.3566	0.1425	−0.2863	1.2809	0.4043
0.6379	−0.4428	−2.3006	−0.6446	0.1516	−1.7708	2.8854
0.4686	1.4664	1.6852	−0.9690	−0.0831	−0.5863	0.8574
−0.5557	0.8115	−0.2676	−1.2496	−1.2125	1.3846	1.1572
0.9990	−0.1032	0.5405	−0.6022	0.0093	0.2119	−1.4647
−0.4428	−0.5564	−0.5098	−1.1929	−0.0572	−0.5061	−0.1557
−1.2384	−0.3924	1.7981	0.6141	−1.3596	1.4943	−0.4406
−0.2033	−0.1316	0.8319	0.4270	−0.8888	0.4167	−0.8513
1.1054	1.2237	−0.7003	0.9780	−0.7679	0.8960	0.5154
−0.7165	0.8563	−1.1630	1.8800			

Again About the Hit-or-Miss Example

Now it is possible to explain how the random points in Figures 1 and 2 are generated. The points in Figure 1 have coordinates

$$x = \gamma', \quad y = \gamma''$$

The values of γ' and γ'' are computed from groups of five digits from Table 1.1: $x_1 = 0.86515$; $y_1 = 0.90795$; $x_2 = 0.66155$; $y_2 = 0.66434$; etc.

Since the abscissas and the ordinates of these points are independent and uniformly distributed over $0 < x < 1$ and $0 < y < 1$, it can be proved that the probability of one such point falling into any region inside the square is equal to the area of the region. In other words, these points are uniformly distributed over the square $0 < x < 1$, $0 < y < 1$.

The points in Figure 2 have coordinates

$$x = 0.5 + 0.2\zeta', \quad y = 0.5 + 0.2\zeta''$$

where the values ζ' and ζ'' are taken successively from Table 1.2: $x_1 = 0.5 + 0.2 \cdot 0.2005$; $y_1 = 0.5 + 0.2 \cdot 1.1922$;

$x_2 = 0.5 + 0.2 \cdot (-0.0077)$; $y_2 = 0.5 + 0.2 \cdot 0.0348$; etc. (One of the points falls outside the square and is rejected.)

As follows from (1.25), the abscissas and ordinates of these points are normal variables with means $a = 0.5$ and variances $\sigma^2 = 0.04$.

Let us estimate the computational error of the hit-or-miss example. The random value N' is approximately normal since N' is a sum

$$N' = \xi_1 + \xi_2 + \ldots + \xi_N$$

where $\xi_j = 1$ if the jth point falls inside S, and $\xi_j = 0$ if otherwise. All these ξ_j are independent and have a common distribution

$$\xi \sim \begin{pmatrix} 1 & 0 \\ S & 1-S \end{pmatrix}$$

here the area of S is denoted by the same letter S. Thus, $\mathbf{M}\xi = S$, $\mathbf{M}\xi^2 = S$, $\mathbf{D}\xi = S - S^2$. The variance of our estimate N'/N is equal to

$$\mathbf{D}(N'/N) = \mathbf{D}\xi/N = S(1-S)/N$$

and its probable error is approximately $0.6745 \times \sqrt{S(1-S)/N}$. At $S = 0.35$, $N = 40$ we obtain the value 0.051 which is in close agreement with the actual error of the computation (0.05).

notes

simulating random variables

examples of the application of the monte carlo method

simulation of a mass-servicing system

Description of the Problem

Consider a simple servicing system that has n "lines" (or "channels", or "distribution points") performing a set of operations that we will call "servicing". The system receives requests arriving at random moments: $T_1 < T_2 < \ldots < T_k < \ldots$.

Let T_k be the moment of arrival of the kth request. If line 1 is free at $t = T_k$, it starts servicing the request; this takes t_h minutes (t_h is the *holding time* of the line). If line 1 is busy at $t = T_k$, the request is immediately transferred to line 2. And so on...

Finally, if all n lines are busy at instant T_k, the system rejects the request.

The problem is to determine how many requests (on average) the system satisfies during an interval of time T; how many rejections will be given?

It is clear that problems of this type are encountered in a variety of organizations ranging from those that provide simple services to those that require highly specialized and complex logistical coordination. In complex situations, like those we shall describe later, the Monte Carlo method turns out to be the only possible method of computation; it is rare that an analytical solution is found.

The Simple Flow of Requests

The first question that arises when such servicing systems are analyzed is: what is the mathematical model of the flow of incoming requests? This question is usually answered by experimental observations of similar systems over long periods of time. Investigation of request flows under various conditions permits us to single out some frequently encountered models.

A *simple flow* (or a *Poisson flow*) is a sequence of events in which the interval between any two consecutive events is an independent random variable with density

$$p(x) = ae^{-ax}, \quad 0 < x < \infty \quad (2.1)$$

This density (2.1) is also called the *exponential distribution* (see Figure 2.1, in which two densities for $a = 1$ and $a = 2$ are constructed).

It is easy to compute the mathematical expectation of a random variable τ with density (2.1):

$$\mathbf{M}\tau = \int_0^\infty xp(x)\,dx = \int_0^\infty xae^{-ax}\,dx$$

Integration by parts ($u = x$, $dv = ae^{-ax}\,dx$) yields

$$\mathbf{M}\tau = \left[-xe^{-ax}\right]_0^\infty + \int_0^\infty e^{-ax}\,dx = \left[-\frac{e^{-ax}}{a}\right]_0^\infty = \frac{1}{a}$$

The parameter a is called the *request flow density*.

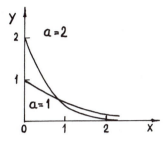

Fig. 2.1. Two exponential densities.

The formula for modeling τ is easily obtained from Equation 1.23, which in our case is written:

$$\int_0^\tau ae^{-ax}\,dx = \gamma$$

Computing the integral on the left, we get the relation

$$1 - e^{-a\tau} = \gamma$$

from which, in turn, we get

$$\tau = -\frac{1}{a}\ln(1-\gamma)$$

However, the variable $1 - \gamma$ has exactly the same distribution as γ, and so, instead of this last equation, we can use the equation

$$\tau = -\frac{1}{a}\ln\gamma \tag{2.2}$$

The Computation Plan

Let us consider the functioning of a system in the case of a simple flow of requests. To each of the n lines we assign one address in RAM, in which we register the moment when this line becomes free. We denote the moment at which the ith line becomes free by t_i.

We begin the calculation at the time when the first request enters the system, so $T_1 = 0$. At this point all the lines are free: $t_1 = t_2 = \ldots = t_n = 0$. The calculation will be finished at time $T_f = T_1 + T$.

The first request enters line 1; this line is now busy for the period t_h, and we must replace t_1 by a new value $(t_1)_{new} = T_1 + t_h$, add one to the counter of serviced requests, and turn to examine the second request.

Let us now assume that k requests have already been considered. It is necessary to select the time of arrival of the $(k+1)$th request. For this we generate the next value of γ and compute the next value $\tau = \tau_k$ using Equation 2.2. Then the entrance time

$$T_{k+1} = T_k + \tau_k$$

Is the first line free at this time? To find out, it is necessary to check the condition

$$t_1 \leq T_{k+1} \tag{2.3}$$

If this condition is met, it means that at time T_{k+1} the line is free and can service the request. We therefore replace t_1 by $T_{k+1} + t_h$, add one to the counter, and turn to the next request.

If condition (2.3) is not satisfied, it means that at moment T_{k+1}, the first line is busy. Then we must test whether the second line is free:

$$t_2 \leq T_{k+1} \tag{2.4}$$

If condition (2.4) is met, we replace t_2 by $T_{k+1} + t_h$, add one to the counter, and turn to the next request.

If condition (2.4) is *not* satisfied, we proceed to test the condition

$$t_3 \leq T_{k+1}$$

It can happen that for all i from 1 to n,

$$t_i > T_{k+1}$$

that is, all lines are busy at time T_{k+1}. We then add one to the counter of rejections and turn to the next request.

Each time T_{k+1} is computed, we have to check the condition for termination of the experiment: $T_{k+1} > T_f$. Once this condition is satisfied, the experiment comes to an end. The counters give us the number of satisfied requests μ_{sat} and the number of rejected requests μ_{rej}.

This experiment must be repeated N times (each time with different values of γ). Then the results of all the trials are averaged:

$$\mathbf{M}\mu_{sat} \approx \frac{1}{N}\sum_{j=1}^{N}\mu_{sat,j}$$

and

$$\mathbf{M}\mu_{rej} \approx \frac{1}{N}\sum_{j=1}^{N}\mu_{rej,j}$$

where the values $\mu_{sat,j}$ and $\mu_{rej,j}$ are the values of μ_{sat} and μ_{rej} obtained in the jth trial.

More Complex Problems

It is easy to see that we can use the same method to compute results for much more complex systems. For example, the holding time t_h may be random and different for the various lines (this would correspond to the use of different servicing equipment or to the employment of service staff having differing qualifications). The scheme of computations remains unchanged, but the values of t_h must be modeled for each request, and the modeling formulas must be specific for each line.

We can consider so-called *systems with waiting*, which do not overflow immediately; a request is stored for some time (its waiting time in the system) and is not rejected if any line becomes available during that time.

Systems can also be considered in which the next request is serviced by the first line that becomes free. We can take into account a random failure of each line and a random time interval necessary to repair it. It is possible to allow for variations in the density of the request flow over time, and in many additional factors.

Of course, a price has to be paid for such simulations. Useful results are obtained only if a sound model is chosen. This requires careful study of actual request flows, timings of the work at the various distribution points, and so forth.

Generally, we must first ascertain the probabilistic laws governing the functioning of the various parts of the system. Then the Monte Carlo method permits the computation of the probabilistic laws of the entire system, however complex it may be.

Such methods of calculations are extremely helpful in planning enterprises. Instead of an expensive (and sometimes impossible) real experiment, we can carry out experiments on a computer, trying out various versions of job organization and equipment usage.

calculating the quality and reliability of devices

The Simplest Scheme for Estimating Quality

Let us consider a device made up of a certain (possibly large) number of elements. For example, this device may be a piece of electrical equipment, made of resistors (R_k), capacitors (C_k), tubes, and the like. We assume that the quality of the device is determined by the value of an output parameter U, which can be computed from the parameters of all the elements:

$$U = f(R_1, R_2, \ldots; C_1, C_2, \ldots; \ldots) \qquad (2.5)$$

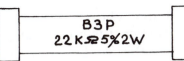

Fig. 2.2. A resistor witn resistance $22 \ k\Omega \pm 5\%$. B3P is the Russian trademark.

If, for example, U is the voltage in an operating section of an electric circuit, then by Ohm's law it is possible to write equations for the circuit and, solving them, to find U.

In reality the parameters of the elements are never exactly equal to their indicated values. For instance, the resistor illustrated in Figure 2.2 can "test out" anywhere between 20.9 and 23.1 kiloohms (B3P is the Russian trademark). The question here is: what effect do deviations of the parameters of all these elements have on the value of U?

We can try to compute the limits of variation of U, taking the "worst" values of the parameters of each element. However, it is not always clear which set of parameter values are the worst. Furthermore, if the total number of elements is large, the limits thus computed are probably highly overestimated: it is very unlikely that all the parameters are simultaneously functioning at their worst.

It is, therefore, more reasonable to consider the parameters of all the elements, and the value U itself, to be random, and to try to estimate the mathematical expectation MU and the variance DU. The value MU is the mean value of U for the aggregate of devices, and DU will show what deviations of U from MU will be encountered in practice. Recall from our discussion of continuous random variables in Chapter 1 that, in general,

$$MU \neq f(MR_1, MR_2, \ldots; \ MC_1, MC_2, \ldots; \ldots)$$

The distribution of U cannot be computed analytically if the function f is even slightly complex; however, sometimes this can be estimated experimentally by examining a large lot of manufactured devices. But such examination is not always possible, and certainly not in the design stage.

Let us try to apply the Monte Carlo method to this problem. To do so, we need to know: (1) the probabilistic characteristics of all the elements, and (2) the function f (more exactly, we must know how to calculate the value of U from any specified values R_1, R_2, ...; C_1, C_2, ...; ...).

The probability distributions of the parameters for each element can be obtained experimentally by examining a large batch of such elements. Quite often these distributions are found to be normal. Therefore, many investigators proceed in the following way: they consider, for example, the resistance of a resistor (pictured in Figure 2.2) to be a normal random variable ρ with mathematical expectation $a = \mathbf{M}\rho = 22$ and with $3\sigma = 1.1$. (According to the rule of three sigmas, it is unlikely to get a value of ρ deviating from $\mathbf{M}\rho$ by 3σ on any one trial.)

The scheme for the computation is simple: first, for each element a random value of its parameter is found; then the value of U is computed according to Equation 2.5. Having repeated this numerical experiment N times (with different random numbers), we obtain values U_1, U_2, ..., U_N and assume that approximately

$$\mathbf{M}U \approx \frac{1}{N} \sum_{j=1}^{N} U_j$$

$$\mathbf{D}U \approx \frac{1}{N-1} \left(\sum_{j=1}^{N} U_j^2 - \frac{1}{N} \Big(\sum_{j=1}^{N} U_j \Big)^2 \right)$$

For large N in the last formula the factor $1/(N-1)$ can be replaced by $1/N$, and then this equation will

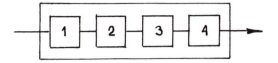

Fig. 2.3. Scheme of a device corresponding to Equation 2.6.

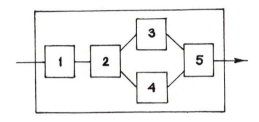

Fig. 2.4. Scheme of a device corresponding to Equation 2.7.

be a direct consequence of Equations 1.8 and 1.9. In statistics it has been shown that for small N it is better to keep the factor $1/(N-1)$.

Examples of the Calculation of Reliability

Suppose we want to estimate how long, on average, a device will function without breakdown, assuming that we know the characteristics of the failure-proof operation of all its components.

If we consider the breakdown time of each component $t_{(k)}$ to be a constant, then the computation of the breakdown time t of the whole device presents no problems. For example, for the device schematically represented in Figure 2.3, in which the failure of one component implies the failure of the entire device,

$$t = \min(t_{(1)}; t_{(2)}; t_{(3)}; t_{(4)}) \qquad (2.6)$$

And for the device in which one of the elements is duplicated with a stand-by element, (schematically represented in Figure 2.4),

$$t = \min\left(t_{(1)}; t_{(2)}; \max(t_{(3)}; t_{(4)}); t_{(5)}\right), \qquad (2.7)$$

since if element 3 fails, for example, the device will continue to work with the remaining element 4.

In practice the duration of failure-proof operation of any component of a device is a random variable. When we say that a light bulb is good for 1,000 hours, we only mean that 1,000 hours is the average value $M\theta$ of the random variable θ. Everyone knows that one bulb may burn out sooner while another will last longer.

If the distribution densities of the breakdown times $\theta_{(k)}$ for all the components of a device are known, $M\theta$ can be computed by the Monte Carlo method, following the plan of the simplest scheme for estimating quality. Indeed, for each component we obtain a value $t_{(k)}$ for the random variable $\theta_{(k)}$. Then, from Equations 2.6 or 2.7, we are able to compute a value t of the random variable θ, representing the breakdown time of the whole device. Repeating this trial enough times (N), we can obtain an approximation

$$M\theta \approx \frac{1}{N} \sum_{j=1}^{N} t_j$$

where t_j is the value of t obtained in the jth numerical experiment.

It should be mentioned that the problem of the probability distributions of breakdown times for individual components is not as simple as one may think: actual experiments are difficult to perform, since one must wait until enough elements have broken down.

Further Possibilities of the Method

The preceding examples show that the techniques for predicting the quality of devices being designed is quite simple in theory. One must know the probabilistic characteristics of all the components of a given

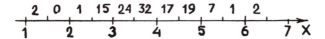

Fig. 2.5. Numbers of values that appeared in each interval.

device, and be able to compute the variable of interest as a function of the parameters of these components. The randomness of the parameters is taken into account by means of our simulation.

From the simulation it is possible to obtain much more information than just the expectation and variance of the variable that interests us. Suppose that we have found a large number of values $U_1, U_2, \ldots, U_N$ of the variable U. From these values we can construct an approximate density of U. (In fact this is a general statistical problem of experimental data processing, though here we have carried out *numerical* experiments.) We are limiting ourselves here to a concrete example.

Let us assume that we have, altogether, 120 values $U_1, U_2, \ldots, U_{120}$ of the random variable U, all contained in the range

$$1.0 < U_j < 6.5$$

Let us divide the interval $1.0 < x < 6.5$ into 11 equal intervals of length $\Delta x = 0.5$ (or any number of intervals that is neither too large nor too small), and count how many values of U_j fall into each interval. The results are given in Figure 2.5.

The frequency of "hits" in any interval is calculated by dividing these numbers by $N = 120$. In our example the frequencies are: 0.017; 0.000; 0.008; 0.120; 0.200; 0.270; 0.140; 0.160; 0.060; 0.008; 0.017.

On each of the intervals of the partition, let us construct a rectangle with area equal to the frequency of hitting that interval (Figure 2.6). In other words, the height of each rectangle is equal to the frequency

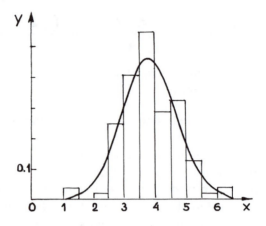

Fig. 2.6. Histogram and density.

divided by Δx. The graph obtained is known as a *histogram*.

The histogram serves as an approximation to the unknown density of the random variable U. Therefore, for example, the area of the histogram between $x = 2.5$ and $x = 5.5$ yields an approximate value of the probability

$$\mathbf{P}\{2.5 < U < 5.5\} \approx 0.95$$

Hence, we can assume on the basis of our experiments that, with a probability of approximately 0.95, the value of U falls in the interval $2.5 < U < 5.5$.

In Figure 2.6 the density of a normal random variable ζ' with the parameters $a = 3.85$, $\sigma = 0.88$ has been constructed as a comparison. If the probability of ζ' falling inside the interval $2.5 < x < 5.5$ is computed from this density, we obtain a fairly close value, 0.91. Indeed, according to Chapter 1 (see Normal Random Variables)

$$\mathbf{P}\{2.5 < \zeta' < 5.5\} = 0.5\big[\Phi(t_2) - \Phi(t_1)\big]$$

where $t_1 = (2.5 - a)/\sigma = -1.54$, $t_2 = (5.5 - a)/\sigma = 1.88$;

therefore,

$$\mathbf{P}\{2.5 < \zeta' < 5.5\} = 0.5\big[\Phi(1.88) + \Phi(1.54)\big] = 0.91$$

A Remark

It is unfortunate that calculations of this type are still fairly scarce — primarily because designers are not aware of this method.

Furthermore, before using the method to evaluate a device, one must find out the probabilistic characteristics of all the components that go into it — this entails a lot of work. Once these characteristics are known, however, one can evaluate the quality of any device that is made of these components. It is even possible to estimate variations in quality when certain components are replaced by others.

One might hope that in the near future such calculations will become routine, and the probabilistic characteristics of different elements will be supplied by their producers.

computation of neutron transmission through a plate

The laws of interaction of single elementary particles (neutrons, photons, mesons, etc.) with matter are known. Usually it is necessary to find out the macroscopic characteristics of processes in which an enormous number of such particles participate: densities, fluxes, and so on. This situation is quite similar to the one we encountered in our discussions of mass servicing systems and the quality and reliability of devices; it, too, can be handled by the Monte Carlo method.

Nuclear physics is, perhaps, the field in which the Monte Carlo method is used most frequently. We will consider the simplest version of the problem of neutron transmission through a plate.

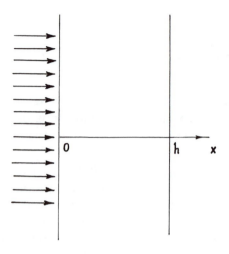

Fig. 2.7. Problem statement.

Statement of the Problem

Let a stream of neutrons with energy E_0 fall on a homogeneous infinite plate $0 \leq x \leq h$, the angle of incidence being $90°$ (Figure 2.7). In collisions with the atoms of which the plate is composed, neutrons can be either elastically scattered or absorbed. Let us assume, for simplicity, that the energy of a neutron does not change in scattering, and that all directions of scattered neutrons are equally probable (this is sometimes the case for neutron collisions with heavy atoms). Figure 2.8 illustrates the possible fates of a neutron: (a) neutron passes through the plate, (b) neutron is absorbed, and (c) neutron is reflected by the plate.

We are required to estimate the probability p^+ of neutron transmission through the plate, the probability p^0 of a neutron being absorbed by the plate, and the probability p^- of neutron reflection from the plate.

Interaction of neutrons with matter is characterized in the case under consideration by two constants

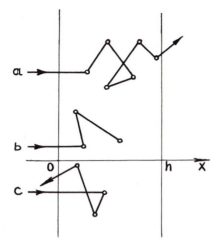

Fig. 2.8. Possible fates of a neutron: (a) transmission, (b) absorption, and (c) reflection.

Σ_a and Σ_s, denoting the *absorption cross-section* and the *scattering cross-section*, respectively. The *total cross-section* in this case is the sum of these two cross-sections:

$$\Sigma = \Sigma_a + \Sigma_s$$

The physical meaning of the cross-sections is as follows: In a collision of a neutron with an atom, the probability of absorption is Σ_a/Σ, and the probability of scattering is Σ_s/Σ.

The distance λ between two consecutive collisions of a neutron is called the *free path length*. This is a random variable; it can assume any positive value with probability density

$$p(x) = \Sigma e^{-x\Sigma}, \quad 0 < x < \infty$$

This density of λ coincides with the density (2.1) of the random variable τ in a simple flow of events (refer to the discussion under The Simple Flow of Requests earlier in this chapter). Accordingly, we can immediately write the expression for the mean free path length

$$M\lambda = 1/\Sigma$$

and an equation for modeling λ:

$$\lambda = -(1/\Sigma)\ln\gamma$$

There remains the question of how to select the random direction of a scattered neutron. Since the situation is symmetric with respect to the x axis, the direction is completely defined by one angle φ between the velocity of a scattered neutron and the x axis. It can be proved (see Transformations of the Type $\xi = g(\gamma)$ in Chapter 3) that the requirement of equal probabilities in each direction is equivalent in this case to the requirement that the cosine of this angle $\mu = \cos\varphi$ be uniformly distributed over the interval $-1 < x < 1$. Equation 1.24 for $a = -1$ and $b = 1$ yields an expression for modeling random values of μ:

$$\mu = 2\gamma - 1$$

Simulation of Real Trajectories

Let us assume that a neutron has undergone its kth scattering inside the plate at a point with abscissa x_k, and afterward began to move in the direction μ_k. We find the free path length

$$\lambda_k = -(1/\Sigma)\ln\gamma$$

and compute the abscissa of the next collision

$$x_{k+1} = x_k + \lambda_k\mu_k$$

(Figure 2.9, in which $\mu_k = \cos\varphi_k$).
We check to see if the condition for passing through the plate has been met:

$$x_{k+1} \geq h$$

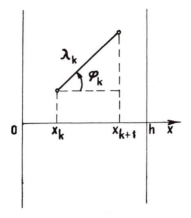

Fig. 2.9. The free path λ_k.

If it has, the computation of the neutron's trajectory stops, and a 1 is added to the counter for transmitted particles. Otherwise, we test the condition for reflection:

$$x_{k+1} \leq 0$$

If this condition is met, the computation of the neutron's trajectory stops, and a 1 is added to the counter for reflected particles. If this condition also fails, that is, if $0 < x_{k+1} < h$, it means that the neutron has undergone its $(k+1)$th collision inside the plate, and we have to determine the fate of the neutron in this collision.

In accordance with the method discussed in Chapter 1 for modeling a discrete random variable, we select the next γ and test the condition for absorption:

$$\gamma < \Sigma_a/\Sigma$$

If this last inequality holds, the trajectory is terminated and a 1 is added to the counter for absorbed particles. If this last inequality does not hold, we assume that the neutron has been scattered at a point with the abscissa x_{k+1}. Then we find a new direction

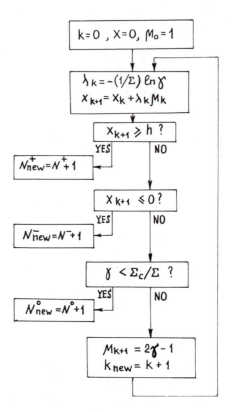

Fig. 2.10. Flowchart for simulating one trajectory.

of movement

$$\mu_{k+1} = 2\gamma - 1$$

and repeat the cycle once more (but, of course, with
new values of γ).

All the γ are written without subscripts, since each
value of γ is used only once. To compute one lap of
the trajectory, three values of γ are needed.

The initial values for every trajectory are

$$x_0 = 0, \quad \mu_0 = 1$$

After N trajectories have been sampled, we find that
N^+ neutrons have gone through the plate, N^- have

been reflected from it, and N^0 have been absorbed. It is obvious that the desired probabilities are approximately equal to the ratios

$$p^+ \approx \frac{N^+}{N}, \; p^- \approx \frac{N^-}{N}, \; \text{and} \; p^0 \approx \frac{N^0}{N}$$

In Figure 2.10 a flowchart of the program for the computation of one trajectory is shown. The subscript k is the collision number along the trajectory.

This computation procedure, although natural, is far from being perfect. In particular, it is difficult to estimate the probability p^+ when it is very small. However, this is precisely the case we encounter in calculating protection against radiation.

Fortunately, there exist more sophisticated versions of the Monte Carlo method that make these computations possible. We shall now briefly discuss one of the simplest methods of calculation with the help of so-called "weights".

Computation Scheme Using Weights that Replace Absorption

Let us reconsider the same problem of neutron transmission. Let us assume that a "package", consisting of a large number w_0 of identical neutrons, is moving along a single trajectory. For a collision at the point with abscissa x_1, the average number of neutrons in the package that would be absorbed is $w_0(\Sigma_a/\Sigma)$, and the average number of neutrons undergoing scattering would be $w_0(\Sigma_s/\Sigma)$. Let us add the value $w_0(\Sigma_a/\Sigma)$ to the counter of absorbed particles, and follow the motion of the scattered package, assuming that all the remaining $w_1 = w_0(\Sigma_s/\Sigma)$ neutrons are scattered in a single direction.

All the equations given for simulation of real trajectories remain the same. But the number of neutrons in the package after each collision is reduced

$$w_{k+1} = w_k(\Sigma_s/\Sigma)$$

since a part of the package, containing $w_k(\Sigma_a/\Sigma)$ neutrons, will be captured. Now the trajectory cannot be terminated by absorption.

The value w_k is usually called the *weight* of the neutron and, instead of talking about a package consisting of w_k neutrons, we speak of a single neutron with weight w_k. The initial weight w_0 is usually set equal to 1. (This is not contrary to what was said about a "large package", since all the w_k obtained while computing a single trajectory contain w_0 as a common factor.)

A flowchart of the program that accomplishes this computation is given in Figure 2.11. Clearly, it is no more complex than the flowchart in Figure 2.10. However, one can prove that the computation of p^+ by the last method is always more precise than the computation by simulation.

Indeed, let us introduce random variables η and η', equal to the number (weight) of neutrons passing through the plate, η and η' are obtained by modeling one trajectory by the method for simulating real trajectories, and by the method for computation using weights that replace absorption, respectively.

If $w_0 = 1$, it is obvious that

$$\mathbf{M}\eta' = \mathbf{M}\eta = p^+$$

(A rigorous proof of this statement can be found in Sobol'.[3])

Since η can assume only two values, 0 and 1, the distribution of η is

$$\eta \sim \begin{pmatrix} 1 & 0 \\ p^+ & 1 - p^+ \end{pmatrix}$$

Taking into account that $\eta^2 = \eta$, we easily find that $\mathbf{D}\eta = p^+ - (p^+)^2$.

It is easy to see that the second variable, η', can assume an infinite sequence of values: $w_0 = 1$, w_1, w_2,

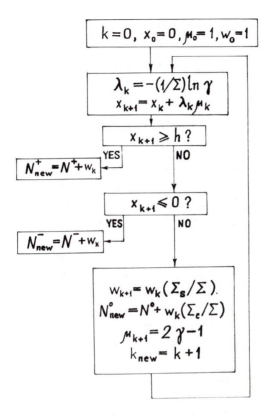

Fig. 2.11. Flowchart for computing one trajectory using weights.

$\ldots, w_k, \ldots$, including the value 0. Therefore its distribution is given by the table

$$\eta' \sim \begin{pmatrix} w_0 & w_1 & w_2 & \ldots & w_k & \ldots & 0 \\ q_0 & q_1 & q_2 & \ldots & q_k & \ldots & q \end{pmatrix}$$

The probabilities q_k need not interest us, since in any case the variance of η' can be written as

$$\mathbf{D}\eta' = \sum_{k=1}^{\infty} w_k^2 q_k - (p^+)^2$$

Finally, noticing that $w_k \leq 1$ and that

$$\sum_{k=1}^{\infty} w_k q_k = \mathbf{M}\eta' = p^+$$

we obtain the inequality $\mathbf{D}\eta' \leq p^+ - (p^+)^2$. Hence, $\mathbf{D}\eta' \leq \mathbf{D}\eta$.

A Remark

There are many ways to perform calculations using weights; we only emphasize that the Monte Carlo method enables us to solve complex problems involving elementary particles: the medium may consist of different substances and can have any geometrical shape; the energy of the particles may change in each collision. Many other nuclear processes can be taken into consideration (for example, the possibility of an atom fission by collision of a neutron, and the subsequent release of new neutrons). Conditions for the initiation and maintenance of a chain reaction can be modeled, and so forth (see An Astrophysical Problem, which follows next in this chapter).

an astrophysical problem

In the next problem the Monte Carlo method is again applied for estimating the transport of particles through matter. Here, the particles are photons; the matter in this case is a spherical plasma cloud.

Comptonization

Since the discovery of X-ray and γ-ray bursts of cosmic origin, astrophysicists have tried to develop models that would explain the observed radiation spectra: the power-law energy distribution of such spectra has been puzzling, and sometimes called "nonthermal".

The simplest model of a compact X-ray source is a cloud of hot plasma with a low-frequency ν photon source at the center. The photon's energy $h\nu$ is increased due to multiple scattering by hot electrons, and it emerges from the cloud as hard X-ray or even γ-ray radiation.

In physics, the change of frequency of a photon that is scattered by an electron is called the Compton effect; the change in the photon spectrum due to multiple scattering of photons by thermal electrons is called the *Comptonization* of radiation. This process is one of the chief mechanisms for producing hard radiation spectra in high-energy astrophysics. Radiation spectra of various compact sources (neutron stars, accretion discs around black holes, quasars, galactic nuclei, etc.) may be considered to be produced by Comptonization. And the most efficient method for modeling such spectra is the Monte Carlo method (see Pozdnyakov et al).[4]

(In reality, the low-energy photons are not necessarily born at a single point: for example, there may be a black hole at the center of the cloud. However, it was proved that the hard-radiation spectrum resulting from Comptonization is insensitive to the spatial distribution of the low-frequency photon source. As physicists say, after several scatterings the photon forgets its birthplace!)

Example: A Powerful Source of X- and γ-Rays

A powerful flux of hard γ-rays coming from the nucleus of the Seyfert galaxy NGC 4151 was detected in 1979. Figure 2.12, from the cited paper (see Pozdnyakov et al),[4] is an attempt to model the observed spectrum of this flux. The flux F_ν is proportional to the probability density of photons escaping from the cloud with energy $h\nu$. The dots are experimental data; error bars (in some measurements rather large) are omitted.

The histogram in Figure 2.12 is computed by the Monte Carlo method.

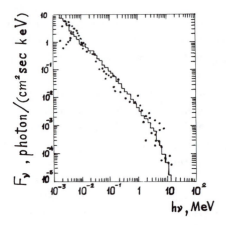

Fig. 2.12. Modeling of the X- and γ-ray spectrum of the nucleus of the Seyfert galaxy NGC 4151. Dots indicate experimental data.

The low-frequency photon source is assumed to have a blackbody (Planck) spectrum with radiation temperature T_r; the hot electrons in the cloud are Maxwellian with temperature T_e, and the cloud is a sphere of radius τ. A satisfactory "fit" is obtained with $kT_r = 0.5$ eV, $kT_e = 2.0$ MeV, and $\tau = 0.4$ in units of photon mean free path with respect to cold electrons.

From Figure 2.12 one can see that over a wide energy range (approximately from $h\nu = 0.001$ MeV up to $h\nu = 6$ MeV), the values of $\lg F_\nu$ are proportional to $\lg h\nu$. From the slope of the histogram one can easily estimate the so-called *spectral index* of the power-law distribution

$$F_\nu \sim (h\nu)^{-1.2}$$

At higher energies ($h\nu > 3kT_e$) there is an exponential cutoff of the spectrum.

On the Method of Computation

As we have already mentioned, the methods for computing neutron transmission through a plate can again be applied in this case. In doing so, the energy of each newly born photon must be modeled according to the Planck distribution; the energy of the scattering electron must be modeled as well; a mean free path for the photon that depends on $h\nu$ must be computed after each scattering; and so forth. We will not tire the reader by reproducing here all the equations involved in these calculations (see Pozdnyakov et al).[4]

It is more interesting that the probabilities on the right end of the spectrum in Figure 2.12 are 10^6 times smaller than the probabilities on the left end. Clearly, these probabilities are not computed by straightforward simulation. The weights used are more sophisticated than those described earlier in the Computation Scheme Using Weights that Replace Absorption. For a photon having weight w_k, after the kth scattering, the probability of direct escape L_k is first computed. Next, a "portion" of the photon with weight $w_k L_k$ is added to the counter of escaping photons, while the "remainder" of the photon with weight $w_{k+1} = w_k(1 - L_k)$ experiences a $(k + 1)$th collision inside the cloud.

evaluating a definite integral

The problems considered in the preceding sections are probabilistic by nature, and to use the Monte Carlo method to solve them seems natural. Now we will consider a purely mathematical problem: the approximate evaluation of a definite integral.

Since evaluating a definite integral is equivalent to calculating an area, the hit-or-miss method discussed in the Introduction could be used. However, we will present here a more efficient method — one that allows us to construct various probabilistic models for solving the problem by the Monte Carlo method. In addition, we will indicate how to select better models from among all alternatives.

The Method of Computation

Let us consider a function $g(x)$ defined over the interval $a < x < b$. Our assignment is to compute approximately the integral

$$I = \int_a^b g(x)\,dx \qquad (2.8)$$

(If the integral is improper, we assume that its convergence is absolute.)

We begin by selecting an arbitrary distribution density $p(x)$ defined over the interval (a, b) — in other words, an arbitrary function $p(x)$ satisfying (1.15) and (1.16). In addition to the random variable ξ (defined over the interval (a, b) with density $p(x)$), we need a random variable

$$\eta = g(\xi)/p(\xi)$$

By (1.18),

$$\mathbf{M}\eta = \int_a^b \frac{g(x)}{p(x)} p(x)\,dx = I$$

Now let us consider N independent, identically distributed random variables $\eta_1, \eta_2, \ldots, \eta_N$ and apply the central limit theorem to their sum. In this case Equation 1.21 is written

$$\mathbf{P}\left\{\left|\frac{1}{N}\sum_{j=1}^{N}\eta_j - I\right| < 3\sqrt{\frac{\mathbf{D}\eta}{\mathbf{N}}}\right\} \approx 0.997$$

This last relation means that if we choose N values $\xi_1, \xi_2, \ldots, \xi_N$, then for sufficiently large N

$$\frac{1}{N}\sum_{j=1}^{N}\frac{g(\xi_j)}{p(\xi_j)} \approx I \qquad (2.9)$$

This form of Equation 1.21 also shows that in all likelihood the error of approximation in (2.9) will not exceed $3(\mathbf{D}\eta/N)^{1/2}$.

Importance Sampling

Again, to compute the integral (2.8), we could use any random variable ξ, defined over the interval (a, b) with density $p(x) > 0$. (In general, the density $p(x)$ may vanish inside (a, b) but only in those points where $g(x) = 0$). In any case

$$\mathbf{M}\eta = \mathbf{M}(g(\xi)/p(\xi)) = I$$

However, the variance $\mathbf{D}\eta$, and, hence, the estimate of this error of approximation depend on what variable ξ we use, since

$$\mathbf{D}\eta = \mathbf{M}(\eta^2) - I^2 = \int\limits_a^b \frac{g^2(x)}{p(x)}\,dx - I^2 \qquad (2.10)$$

Let us prove that $\mathbf{D}\eta$ is minimized when the density $p(x)$ is proportional to $|g(x)|$.

We will use an inequality well known in analysis:

$$\left(\int\limits_a^b |u(x)v(x)|\,dx\right)^2 \le \int\limits_a^b u^2(x)\,dx \int\limits_a^b v^2(x)\,dx$$

If we set $u = g(x)/\sqrt{p(x)}$ and $v = \sqrt{p(x)}$, then we obtain

$$\left(\int\limits_a^b |g(x)|\,dx\right)^2 \le \int\limits_a^b \frac{g^2(x)}{p(x)}\,dx \int\limits_a^b p(x)\,dx$$

$$= \int\limits_a^b \frac{g^2(x)}{p(x)}\,dx \qquad (2.11)$$

It follows from (2.10) and (2.11) that

$$\mathbf{D}\eta \ge \left(\int\limits_a^b |g(x)|\,dx\right)^2 - I^2 \qquad (2.12)$$

Now it remains to prove that the lower bound (2.12) of the variance $\mathbf{D}\eta$ is attained when the density is proportional to $|g(x)|$, that is, when $p = c|g(x)|$. Since the density must satisfy (1.16), it follows that

$$c = \left(\int\limits_a^b |g(x)|\, dx \right)^{-1}$$

Hence,

$$\int\limits_a^b \frac{g^2(x)}{p(x)}\, dx = \frac{1}{c} \int\limits_a^b |g(x)|\, dx = \left(\int\limits_a^b |g(x)|\, dx \right)^2$$

and the right-hand side of (2.10) is indeed equal to the right-hand side of (2.12).

In practice, it is impossible to use the "best" density. To obtain the "best density", we must ascertain the value of the integral $\int\limits_a^b |g(x)|\, dx$; however, evaluation of the last integral is just as difficult as evaluation of the integral (2.8). Therefore, we restrict ourselves to the following suggestion: *It is desirable that $p(x)$ be proportional to $|g(x)|$.*

Of course, very complex $p(x)$ should not be selected, since the modeling of ξ would become excessively laborious. But it is possible to use the suggestion given above as a guide in choosing $p(x)$ (compare this with the following numerical example).

Estimate (2.9), with a density similar to $|g(x)|$, is called *importance sampling.*

We recall that in practice, one-dimensional integrals of the form (2.8) are not computed by the Monte Carlo method — quadrature formulas provide greater accuracy for calculating such integrals. But the situation changes when we turn to multidimensional integrals: quadrature formulas become very complex or inefficient, while the Monte Carlo method remains almost unchanged.

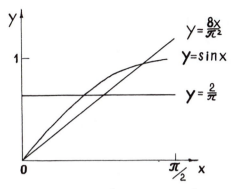

Fig. 2.13. The integrand $y = \sin x$ and two densities.

A Numerical Example

Let us approximately compute the integral

$$I = \int_0^{\pi/2} \sin x \, dx$$

The exact value of this integral is known:

$$\int_0^{\pi/2} \sin x \, dx = \left[-\cos x \right]_0^{\pi/2} = 1$$

We shall use two different random variables ξ for the calculation: one with constant density $p(x) \equiv 2/\pi$ (that is, a uniform distribution in the interval $0 < x < \pi/2$), and the other with linear density $p(x) = 8x/\pi^2$. Both of these densities, together with the integrand $\sin x$, are shown in Figure 2.13. It is evident that the linear density agrees better with the recommendation of importance sampling, that it is desirable for $p(x)$ to be proportional to $\sin x$. Thus one may expect that the second approach will yield the better result.

first approach

Let $p(x) = 2/\pi$ for all x in $(0, \pi/2)$. The formula for modeling ξ can be obtained from (1.24) for $a = 0$ and $b = \pi/2$:

$$\xi = \frac{\pi}{2}\gamma$$

Now Equation 2.9 takes the form

$$I \approx \frac{\pi}{2N} \sum_{j=1}^{N} \sin \xi_j$$

Let $N = 10$. For values of γ let us use groups of three digits from Table 1.1, multiplied by 0.001. The intermediate results are listed in Table 2.1. The final result of the computation is

$$I \approx 0.952$$

second approach

Now let $p(x) = 8x/\pi^2$. For the modeling of ξ we use Equation 1.23:

$$\int_0^\xi \frac{8x}{\pi^2}\, dx = \gamma$$

whence, after some simple calculations, we obtain

$$\xi = \frac{\pi}{2}\sqrt{\gamma}$$

Equation 2.9 takes on the form

$$I \approx \frac{\pi^2}{8N} \sum_{j=1}^{N} \frac{\sin \xi_j}{\xi_j}$$

Let $N = 10$. We use the same random numbers γ as in the first approach. The intermediate results are listed in Table 2.1. The final result of the computation is

$$I \approx 1.016$$

Table 2.1 Intermediate Results

		First approach		Second approach	
j	γ_j	ξ_j	$\sin \xi_j$	ξ_j	$\dfrac{\sin \xi_j}{\xi_j}$
1	0.865	1.359	0.978	1.461	0.680
2	0.159	0.250	0.247	0.626	0.936
3	0.079	0.124	0.124	0.442	0.968
4	0.566	0.889	0.776	1.182	0.783
5	0.155	0.243	0.241	0.618	0.937
6	0.664	1.043	0.864	1.280	0.748
7	0.345	0.542	0.516	0.923	0.863
8	0.655	1.029	0.857	1.271	0.751
9	0.812	1.275	0.957	1.415	0.698
10	0.332	0.521	0.498	0.905	0.868

As we anticipated, the second approach gives the more accurate result.

error estimates

From the values given in Table 2.1, one can approximate the variances $\mathbf{D}\eta$ for both cases. (The computation equation can be found under The Simplest Scheme for Estimating Quality earlier in this chapter.) For the first approach

$$\mathbf{D}\eta \approx \frac{\pi^2}{9 \cdot 4} \left(\sum_{j=1}^{10} (\sin \xi_j)^2 - \frac{1}{10} \left(\sum_{j=1}^{10} \sin \xi_j \right)^2 \right)$$

$$= \frac{\pi^2}{36} (4.604 - 3.670) = 0.256$$

For the second approach

$$\mathbf{D}\eta \approx \frac{\pi^4}{9 \cdot 64} \left(\sum_{j=1}^{10} \left(\frac{\sin \xi_j}{\xi_j} \right)^2 - \frac{1}{10} \left(\sum_{j=1}^{10} \frac{\sin \xi_j}{\xi_j} \right)^2 \right)$$

$$= \frac{\pi^4}{576}(6.875 - 6.777) = 0.016$$

Though the sample size $N = 10$ is small and the applicability of the central limit theorem cannot be guaranteed, let us estimate the probable errors $r_N = 0.6745(\mathbf{D}\eta/N)^{1/2}$ for both computations. For the first approach, $r_N = 0.103$, while for the second, $r_N = 0.027$. Clearly, the actual absolute errors — 0.048 and 0.016, respectively — are of the same order of magnitude as these probable errors.

The exact values of the variances $\mathbf{D}\eta$ are 0.233 and 0.0166. One can see that the second approach, which includes importance sampling, is also more accurate for estimating the variance.

notes

additional information

on pseudorandom numbers

Most of the algorithms for generating pseudorandom numbers have the form

$$\gamma_k = \Phi(\gamma_{k-1}) \text{ for } k = 1,\ 2,\ \ldots \qquad (3.1)$$

If the initial number γ_0 is fixed, all the successive numbers $\gamma_1,\ \gamma_2,\ \ldots$ are calculated by the same Equation 3.1. The mid-square method (see Pseudorandom Numbers in Chapter 1) also has the form of (3.1): a set of operations is given that is to be applied to the argument x to obtain the value y (rather than the analytical expression of the function $y = \Phi(x)$).

The Basic Difficulty of Selecting $\Phi(x)$

Let us prove that the function $\Phi(x)$ shown in Figure 3.1 cannot be used for generating pseudorandom numbers by means of Equation 3.1.

Indeed, we may consider a sequence of points with Cartesian coordinates

$$(\gamma_1,\ \gamma_2),\ (\gamma_3,\ \gamma_4),\ (\gamma_5,\ \gamma_6),\ \ldots$$

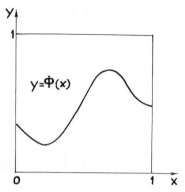

Fig. 3.1. A function that should not be used for generating pseudorandom numbers.

inside the unit square $0 < x < 1$, $0 < y < 1$. Since here we have $\gamma_2 = \Phi(\gamma_1)$, $\gamma_4 = \Phi(\gamma_3)$, $\gamma_6 = \Phi(\gamma_5),\ldots$, all these points are located on the curve $y = \Phi(x)$. This, of course, is unacceptable, because true random points must uniformly fill the whole square.

Thus, we conclude that a function $y = \Phi(x)$ can be successfully used in Equation 3.1 only if its graph provides a sufficiently dense filling of the square!

An example of a function possessing this property is

$$y = \{gx\} \tag{3.2}$$

where g is a very large integer, and $\{z\}$ denotes the fractional part of z. Function (3.2) is plotted in Figure 3.2 for $g = 20$. (The reader can imagine what this graph looks like at $g = 5^{17}$.)

The Congruential Method

The most popular algorithm for generating pseudorandom numbers was suggested by D. H. Lehmer in 1949. This algorithm is based on Equation 3.2, but to avoid round-off errors, the calculation formulas are written in a different form.

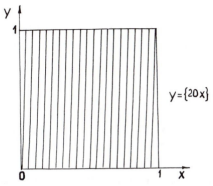

Fig. 3.2. A function whose graph fills the square.

A sequence of integers m_k is defined in which the initial number m_0 is fixed, and all subsequent numbers m_1, m_2, ... are computed by one formula:

$$m_k \equiv gm_{k-1}(\text{mod } M) \qquad (3.3)$$

for $k = 1, 2, \ldots$; from the numbers m_k we calculate the pseudorandom numbers

$$\gamma_k = m_k/M \qquad (3.4)$$

The integer g is called *multiplier*, and the integer M is referred to as the *modulus*.

Relation (3.3) means that the number m_k is equal to the remainder left when gm_{k-1} is divided by M. In the theory of congruence (see any textbook on number theory) this remainder is known as the *least positive residue modulo M*. Triplets (g, M, m_0) can be found that produce relatively satisfactory sequences of pseudorandom numbers.

Example 1

Let $g = 5^{17}$, $M = 2^{40}$, $m_0 = 1$. This generator has been successfully used in Russian computers operating with 40-bit numbers. Here both formulas (3.3) and (3.4) are easily realized: carry out a double precision multiplication $g \cdot m_{k-1}$, and use the low-order digits of the product. The period of the sequence m_k is $P = 2^{38} \approx 2.7 \cdot 10^{11}$.

Example 2

Let $g = 65539 = 2^{16} + 3, M = 2^{29}, m_0 = 1$. This is the infamous RANDU generator, supplied by IBM in their 360 series and included in the Russian ES computers' software. Many users have obtained unsatisfactory results in computations using pseudorandom numbers produced by the RANDU generator. A clear proof that this multiplier is inadequate can be found in Forsythe et al.[5] (Nevertheless, there are still attempts to carry out computations using RANDU.)

Pseudorandom Numbers for Personal Computers

Ordinary congruential generators are not suitable for computers with short words: if the modulus is small, the period is small; if a long word generator is implemented "by parts", generation is slow.

It was proposed by Wichman and Hill to use in parallel three very short word generators,[6]

$$m_k \equiv 171m_{k-1}(\text{mod}\,30269)$$

$$m'_k \equiv 172m'_{k-1}(\text{mod}\,30307)$$

$$m''_k \equiv 170m''_{k-1}(\text{mod}\,30323)$$

and to consider as pseudorandom numbers the fractional parts

$$\gamma_k = \left\{ \frac{m_k}{30269} + \frac{m'_k}{30307} + \frac{m''_k}{30323} \right\}$$

This algorithm can be used in any computer with a word length of 16 (or more) bits. The period of the sequence γ_k is $P \approx 6.9 \cdot 10^{12}$.

Numerical experiments by Levitan and Sobol'[7] show that, contrary to the suggestion of Wichman and Hill,[6] the initial values m_0, m_0', m_0'' should not be selected at random. A good sequence can be obtained with $m_0 = 5$, $m_0' = 11$, $m_0'' = 17$ (almost ten million γ_k were tested).

Program

```
FUNCTION RANDOM()
C    RETURNS A PSEUDO RANDOM NUMBER UNIFORMLY
C    DISTRIBUTED BETWEEN 0 AND 1
C    SET IX=5, IY=11, IZ=17 BEFORE FIRST ENTRY
C    INTEGER ARITHMETIC TO 30323 REQUIRED
        COMMON/RAND/IX, IY,IZ
        IX=171*MOD(IX, 177)-2*(IX/177)
        IY=172*MOD(IY, 176)-35*(IY/176)
        IZ=170*MOD(IZ, 178)-63*(IZ/178)
C
        IF(IX.LT.0)IX=IX+30269
        IF(IY.LT.0)IY=IY+30307
        IF(IZ.LT.0)IZ=IZ+30323
C
        RANDOM=MOD(IX/30269.+IY/30307.+IZ/30323.,1.)
        RETURN
        END
```

---

on methods for generating random variables

This section contains a brief description of the most important transformations used for modeling random variables. These transformations are classified according to the quantities of random numbers that are used for obtaining a single random value. This classification was introduced by Sobol',[3] and is of crucial importance for quasi-Monte Carlo methods (see Constructive Dimension of a Monte Carlo Algorithm below).

Transformations of the Type $\xi = g(\gamma)$

The most important transformation of this type is the *method of inverse functions* (sometimes called the *direct method*). We will now show that methods of modeling discrete and continuous random variables described in Chapter 1 (see Transformations of Random Variables) are special cases of the general inverse functions method.

First, let us recall that the *probability distribution function* $F(x)$ of an arbitrary random variable ξ (or the *cumulative distribution function*), is defined for all $-\infty < x < \infty$ by the relation

$$F(x) = \mathbf{P}\{\xi < x\}$$

(refer to any textbook on probability theory). It is a non-decreasing function, and both limits exist:

$$\lim_{x \to -\infty} F(x) = 0$$

and

$$\lim_{x \to \infty} F(x) = 1$$

However, $F(x)$ is not necessarily strictly monotonic: there may be intervals of constancy. Also, it is not necessarily continuous — there may be jumps.

Second, assume that the function $y = F(x)$ is continuous and strictly monotonic (Figure 3.3). Then a continuous inverse function $x = G(y)$ exists, and for all $-\infty < x < \infty$ and $0 < y < 1$

$$G(F(x)) = x, \; F(G(y)) = y \tag{3.5}$$

It is easy to prove that the distribution function of $G(\gamma)$ is equal to $F(x)$. Clearly,

$$\mathbf{P}\{G(\gamma) < x\} = \mathbf{P}\{F(G(\gamma)) < F(x)\}$$
$$= \mathbf{P}\{\gamma < F(x)\}$$

Since γ is uniformly distributed over the unit interval, the probability

$$\mathbf{P}\{\gamma < F(x)\} = \mathbf{P}\{0 < \gamma < F(x)\}$$

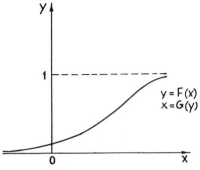

Fig. 3.3. Continuous and strictly monotonic distribution function $y = F(x)$ and its inverse.

is equal to the length of the interval $(0, F(x))$; so,

$$\mathbf{P}\{G(\gamma) < x\} = F(x)$$

Hence, a random variable ξ having such a probability distribution function $F(x)$ can be modeled by the equation

$$\xi = G(\gamma) \tag{3.6}$$

However, the variable $1 - \gamma$ is uniformly distributed over the same interval as γ. Therefore the formula

$$\xi = G(1 - \gamma) \tag{3.7}$$

can be used instead of (3.6).

Both transformations (3.6) and (3.7) represent the *method of inverse functions*.

The General Method

Consider now an arbitrary random variable ξ with an arbitrary distribution function $F(x)$, so that a classical inverse function $G(y)$ may be multivalued or not defined for all $0 < y < 1$. In this case a generalized inverse function can be introduced that does not sat-

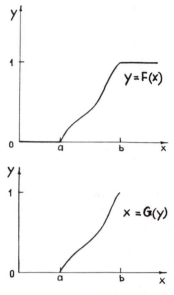

Fig. 3.4. Distribution function $y = F(x)$ of a continuous random variable and its inverse.

isfy (3.5), but that does satisfy a weaker requirement: that inequalities $G(y) \geq x$ and $y \geq F(x)$ be equivalent. Then the previous proof can be slightly modified:

$$\mathbf{P}\{G(\gamma) < x\} = 1 - \mathbf{P}\{G(\gamma) \geq x\}$$

$$= 1 - \mathbf{P}\{\gamma \geq F(x)\} = \mathbf{P}\{\gamma < F(x)\} = F(x)$$

Example 1

Consider the continuous random variable ξ from Modeling of a Continuous Random Variable in Chapter 1. The curve shown in Figures 1.9 and 1.10 is the nontrivial part of the distribution function

$$F(x) = \mathbf{P}\{\xi < x\} = \mathbf{P}\{a < \xi < x\} \cdot$$
$$= \int_a^x p(x)\, dx \text{ for } a < x < b \tag{3.8}$$

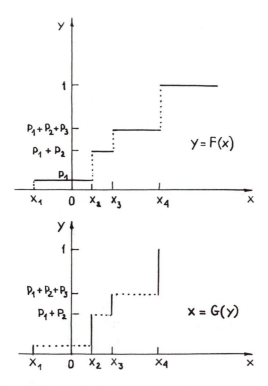

Fig. 3.5. Distribution function $y = F(x)$ of a discrete random variable and its inverse.

The full distribution function $y = F(x)$ is shown in Figure 3.4, together with the inverse function $x = G(y)$. Here the "generalization" is very simple: we exclude the multivalued tails at $x < a$ and $x > b$. Clearly, Equation 1.23 is identical to Equation 3.6.

Example 2

Consider the discrete random variable ξ from Modeling of a Discrete Random Variable, in Chapter 1. Its distribution function $y = F(x)$, together with the generalized inverse function $x = G(y)$, can be seen in Figure 3.5 (for the special case $n = 4$). If we select a

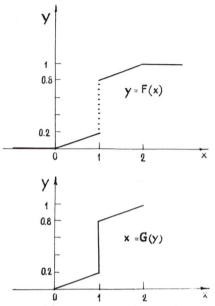

Fig. 3.6. Distribution function $y = F(x)$ of a mixed random variable and its inverse.

value $y = \gamma$ on the y axis, the corresponding $G(\gamma)$ is equal to one of the x_i, and

$$\mathbf{P}\{G(\gamma) = x_i\} = p_i$$

The "generalization" here is somewhat different: we add vertical bars (so that the graph becomes continuous) and then exclude the horizontal bars (so that the function $G(y)$ becomes univalent).

Example 3

Now consider a random variable ξ of a mixed type: $\xi = 1$, with probability 0.6, and it is uniformly distributed in the interval $0 < x < 2$ with probability 0.4. In Figure 3.6 the discontinuous distribution function $y = F(x)$ and the generalized inverse function $x = G(y)$ are shown.

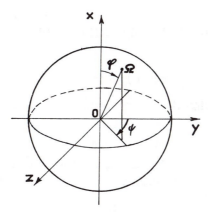

Fig. 3.7. A random point on a sphere and its spherical coordinates.

Since

$$F(x) = \begin{cases} 0.2x & \text{for } 0 < x \le 1 \\ 0.6 + 0.2x & \text{for } 1 < x \le 2 \end{cases}$$

the formula for modeling ξ according to (3.6) is

$$\xi = \begin{cases} 5\gamma & \text{if } 0 < \gamma < 0.2 \\ 1 & \text{if } 0.2 < \gamma < 0.8 \\ 5(\gamma - 0.6) & \text{if } 0.8 < \gamma < 1 \end{cases}$$

a random direction in space

The direction may be specified by a unit vector from the origin whose endpoint lies on the surface of the unit sphere. The words "all directions are equally probable" mean that the endpoint Ω is a random point uniformly distributed on the surface of the sphere. The probability that Ω belongs to an element of the surface dS is equal to $dS/(4\pi)$.

Consider spherical coordinates (φ, ψ) on the sphere with the polar axis Ox (Figure 3.7). Then

$$dS = \sin \varphi \, d\varphi \, d\psi$$

where $0 \leq \varphi \leq \pi$, $0 \leq \psi < 2\pi$. Let us denote by $p(\varphi, \psi)$ the probability density of the point $\Omega = (\varphi, \psi)$. The requirement

$$p(\varphi, \psi) \, d\varphi \, d\psi = dS/(4\pi)$$

together with the last expression for dS, yield the relation

$$p(\varphi, \psi) = (4\pi)^{-1} \sin \varphi$$

Marginal densities of φ and ψ can be found easily from their joint density:

$$p_1(\varphi) = \int\limits_0^{2\pi} p(\varphi, \psi) \, d\psi = \tfrac{1}{2} \sin \varphi$$

$$p_2(\psi) = \int\limits_0^\pi p(\varphi, \psi) \, d\varphi = \tfrac{1}{2\pi}$$

From the identity $p(\varphi, \psi) = p_1(\varphi)p_2(\psi)$ we conclude that φ and ψ are independent and may be modeled independently using two random numbers.

First, we apply the method of inverse functions for modeling φ. Since for $0 \leq \varphi \leq \pi$

$$F(\varphi) = \int\limits_0^\varphi p_1(\varphi) \, d\varphi = \frac{1}{2}(1 - \cos \varphi)$$

we may use (3.7):

$$\frac{1}{2}(1 - \cos \varphi) = 1 - \gamma$$

from which it follows that

$$\cos \varphi = 2\gamma - 1 \qquad\qquad (3.9)$$

Then we apply the same method (in fact, Equation 3.6) for modeling ψ that is uniformly distributed between 0 and 2π:

$$\psi = 2\pi\gamma \qquad\qquad (3.10)$$

Equations 3.9 and 3.10 make possible the choice of a random (isotropic) direction in space. Of course, the values of γ in (3.9) and (3.10) are two different random numbers.

Transformations of the Type $\xi = g(\gamma_1, \gamma_2)$

The most important transformation of this type is the *superposition* (or *composition*) method.

Assume that the probability distribution function $F(x)$ can be regarded as a superposition of several distribution functions $F_1(x), \ldots, F_m(x)$:

$$F(x) = \sum_{k=1}^{m} c_k F_k(x) \qquad (3.11)$$

with all $c_k > 0$ and $c_1 + \ldots + c_m = 1$. Let us assume further that random variables with distribution functions $F_k(x)$ can be modeled, for example, using inverse functions $G_k(y)$.

Then we may introduce a random integer $\varkappa$ with distribution

$$\varkappa \sim \begin{pmatrix} 1 & 2 & \ldots & m \\ c_1 & c_2 & \ldots & c_m \end{pmatrix} \qquad (3.12)$$

so that $\mathbf{P}\{\varkappa = k\} = c_k$, now we can define a two-stage modeling procedure: selecting two random numbers γ_1 and γ_2, 1) use γ_1 to define a random value $\varkappa = k$ and 2) use γ_2 to define $\xi = G_k(\gamma_2)$. The distribution function of ξ is $F(x)$.

proof

We may use a well-known expression for the absolute probability:

$$\mathbf{P}\left\{ G_\varkappa(\gamma_2) < x \right\}$$

$$= \sum_{k=1}^{m} \mathbf{P}\left\{ G_\varkappa(\gamma_2) < x | \varkappa = k \right\} \mathbf{P}\left\{ \varkappa = k \right\}$$

The conditional probabilities here are evident:

$$\mathbf{P}\left\{ G_\varkappa(\gamma_2) < x | \varkappa = k \right\} = \mathbf{P}\{ G_k(\gamma_2) < x \} = F_k(x)$$

Fig. 3.8. Scattering angle θ.

Hence,

$$\mathbf{P}\left\{G_{x}(\gamma_{2}) < x\right\} = \sum_{k=1}^{m} F_{k}(x)c_{k} = F(x)$$

Clearly, if all corresponding densities exist, we may consider the superposition of densities rather than (3.11):

$$p(x) = \sum_{k=1}^{m} c_{k}p_{k}(x)$$

Example 1

When photons are scattered by cool electrons, the scattering angle θ (Figure 3.8) is a random variable, and its cosine $\mu = \cos\theta$ obeys Rayleigh's law:

$$p(x) = \frac{3}{8}(1 + x^2) \quad \text{for } -1 < x < 1$$

An attempt to apply the method of inverse functions for modeling μ yields a cubic equation

$$\frac{1}{8}\left(\mu^3 + 3\mu + 4\right) = \gamma$$

As an alternative, let us consider a representation of the density

$$p(x) = 0.75\, p_1(x) + 0.25\, p_2(x)$$

with $p_1(x) \equiv 0.5$, which is a constant density, and $p_2(x) = 1.5x^2$. The corresponding distribution functions are quite simple:

$$F_1(x) = \frac{1}{2}(x + 1)$$

additional information

and
$$F_2(x) = \frac{1}{2}(x^3 + 1)$$

In both cases we can use inverse functions for obtaining a final explicit formula:

$$\mu = \begin{cases} 2\gamma_2 - 1 & \text{if } \gamma_1 < 0.75 \\ (2\gamma_2 - 1)^{1/3} & \text{if } \gamma_1 > 0.75 \end{cases}$$

Example 2

Consider a discrete random variable

$$\xi \sim \begin{pmatrix} x_1 & \cdots & x_n \\ p_1 & \cdots & p_n \end{pmatrix} \tag{3.13}$$

where all the probabilities p_i are of the form $p_i = m_i 2^{-s}$ with integers m_i, $1 \le m_i \le 2^s - 1$. Assume that s is much smaller than n. Then it may be expedient to regard ξ as a superposition of several (not more than s) random variables with equiprobable values, since we know that such variables can be modeled easily. We will illustrate this approach with a numerical example.

Let the number of possible values in (3.13) be $n = 19$, and all the probabilities $p_i = m_i/64$; the numerators m_i are given in Table 3.1. On the right-hand side of Table 3.1 the same m_i are written in the binary system; v_k is the number of ones in the kth column.

Hence it follows that ξ can be represented in the form of a superposition of three random variables $\xi^{(k)}$, with $k = 4$, 5, 6. The variable $\xi^{(4)}$ assumes values $x_1 - x_8$ with probability 1/8; the variable $\xi^{(5)}$ assumes values x_1, x_2, $x_9 - x_{16}$ with probability 1/10; the variable $\xi^{(6)}$ assumes values $x_3 - x_6$, $x_9 - x_{13}$, $x_{17} - x_{19}$, with probability 1/12. The corresponding coefficients c_k can be computed from the relation $c_k = v_k 2^{-k}$, and are equal to $c_4 = 1/2$, $c_5 = 5/16$, $c_6 = 3/16$.

Denote

$$(y_1, y_2, \ldots, y_{10}) = (x_1, x_2, x_9, x_{10}, \ldots, x_{16})$$

Table 3.1. Numerators m_i

i	m_i	k=1	2	3	4	5	6
1	6				1	1	0
2	6				1	1	0
3	5				1	0	1
4	5				1	0	1
5	5				1	0	1
6	5				1	0	1
7	4				1	0	0
8	4				1	0	0
9	3					1	1
10	3					1	1
11	3					1	1
12	3					1	1
13	3					1	1
14	2					1	0
15	2					1	0
16	2					1	0
17	1						1
18	1						1
19	1						1
v_k					8	10	12

and

$$(z_1, \ z_2, \ \ldots, \ z_{12}) = (x_3, \ x_4, \ x_5, \ x_6, \ x_9, \ x_{10},$$

$$\ldots, \ x_{13}, \ x_{17}, \ x_{18}, \ x_{19})$$

Then the final formula for modeling ξ can be written in the following form:

$$\xi = \begin{cases} x_i & i = 1 + [8\gamma_2] & \text{if } \gamma_1 < 1/2 \\ y_i & i = 1 + [10\gamma_2] & \text{if } 1/2 < \gamma_1 < 13/16 \\ z_i & i = 1 + [12\gamma_2] & \text{if } 13/16 < \gamma_1 \end{cases}$$

In this example the general method for modeling a discrete random variable would require many comparisons of γ with $p_1, \ p_1 + p_2, \ p_1 + p_2 + p_3, \ \ldots$

modeling of a normal random variable

Consider a normal random variable ξ, with $a = 0$ and $\sigma = 1$. Let η be a random variable with the same distribution, but independent of ξ. Then the probability density of a random point with Cartesian coordinates $(\xi, \ \eta)$ in the $(x, \ y)$ plane is equal to the product of one-dimensional densities:

$$p(x, \ y) = \frac{1}{\sqrt{2\pi}} e^{-x^2/2} \frac{1}{\sqrt{2\pi}} e^{-y^2/2} = \frac{1}{2\pi} e^{-(x^2+y^2)/2}$$

Let us introduce polar coordinates

$$x = r \cos \varphi \text{ and } y = r \sin \varphi$$

and denote by ρ, θ the polar coordinates of the point $(\xi, \ \eta)$:

$$\xi = \rho \cos \theta, \quad \eta = \rho \sin \theta$$

The joint density of ρ and θ can be computed easily:

$$\tilde{p}(r, \ \varphi) = p(x, \ y) \left| \frac{\partial(x, \ y)}{\partial(r, \ \varphi)} \right| = \frac{r}{2\pi} e^{-r^2/2}$$

Then the individual densities of ρ and θ can be found by integration:

$$p_1(r) = \int_0^{2\pi} p(r, \ \varphi) \, d\varphi = r \, e^{-r^2/2}$$

and

$$p_2(\varphi) = \int_0^\infty p(r, \varphi) \, dr = \frac{1}{2\pi}$$

Since $p_1(r)p_2(\varphi) \equiv \tilde{p}(r, \varphi)$, we conclude that ρ and θ are independent and can be modeled independently using the corresponding distribution functions

$$F_1(r) = 1 - e^{-r^2/2}$$

and

$$F_2(\varphi) = \varphi/(2\pi)$$

here $0 < r < \infty$, $0 < \varphi < 2\pi$.

From the equations

$$F_1(\rho) = 1 - \gamma_1$$

and

$$F_2(\theta) = \gamma_2$$

we obtain explicit expressions

$$\rho = (-2\ln \gamma_1)^{1/2}, \quad \varphi = 2\pi\gamma_2$$

The final formulas

$$\xi = (-2\ln \gamma_1)^{1/2} \cos 2\pi\gamma_2$$

and

$$\eta = (-2\ln \gamma_1)^{1/2} \sin 2\pi\gamma_2$$

allow the computation of two independent normal variates (with $a = 0$ and $\sigma = 1$) from two random numbers γ_1 and γ_2. However, each of these formulas is of the type $\xi = g(\gamma_1, \gamma_2)$.

Transformations of the Type $\xi = g(\gamma_1, \ldots, \gamma_n)$

Example 1

The random variable ξ with distribution function $F(x) = x^n$ for $0 < x < 1$ can be modeled by the equation

$$\xi = \max(\gamma_1; \ldots; \gamma_n) \tag{3.14}$$

Indeed, the probability

$$P\{\xi < x\} = P\left\{\max_{1 \le i \le n} \gamma_i < x\right\}$$

$$= P\{\gamma_1 < x, \ldots, \gamma_n < x\}$$

and, since all the random numbers are independent,

$$P\{\gamma_1 < x, \ldots, \gamma_n < x\}$$

$$= P\{\gamma_1 < x\} \ldots P\{\gamma_n < x\} = x^n$$

The same random variable ξ can be modeled by the method of inverse functions (3.6). Then the equation $\xi^n = \gamma$ yields the expression

$$\xi = \sqrt[n]{\gamma} \qquad (3.15)$$

Comparing (3.14) with (3.15) we conclude that, rather than extract the nth root of a random number, one may select the greatest among n random numbers.

Example 2

In Chapter 2 the simple flow of events was considered. It was characterized by independent exponentially distributed time intervals τ between two successive events. In more general flows, called *Erlang flows*, the time intervals τ are independent random variables with density

$$p(x) = \frac{a^n}{(n-1)!} x^{n-1} e^{-ax} \text{ for } 0 < x < \infty \qquad (3.16)$$

Since the integral

$$\Gamma(n) = \int_0^\infty x^{n-1} e^{-x} \, dx = (n-1)!$$

is called the *gamma-function*, the density (3.16) is called the *gamma-distribution*.

It can be proved that a random variable ξ, with density (3.16), may be modeled by the equation

$$\xi = -\frac{1}{a} \ln \left(\gamma_1 \ \cdots \ \gamma_n \right)$$

We omit the proof (by mathematical induction), and mention only that Equation 2.2 is a special case of the last equation, namely, when $n = 1$.

the use of order statistics

Assume that n random numbers $\gamma_1, \ldots, \gamma_n$ are re-ordered

$$\gamma_{(1)} \leq \gamma_{(2)} \leq \cdots \leq \gamma_{(n)}$$

The value $\gamma_{(s)}$ is called the sth *order statistic* of the uniform distribution. Clearly $\gamma_{(s)}$ depends on all $\gamma_1, \ldots, \gamma_n$. The density of $\gamma_{(s)}$ is

$$p(x) = n \binom{n-1}{s-1} x^{s-1} (1-x)^{n-s}$$

$$0 < x < 1 \tag{3.17}$$

sketch of a proof

We fix an arbitrary interval $(x, \ x + \Delta x)$ inside the unit interval. When a value of γ is selected, one of three events may occur:

$$A_1 = \{\gamma < x\}, \ A_2 = \{x \leq \gamma < x + \Delta x\},$$

$$A_3 = \{x + \Delta x \leq \gamma\}$$

The probabilities of these events are

$$p_1 = x, \ p_2 = \Delta x$$

and

$$p_3 = 1 - x - \Delta x$$

Consider now a sequence of n independent values $\gamma_1, \ldots, \gamma_n$. In these n trials event A_i will occur ν_i

times, so that $\nu_1 + \nu_2 + \nu_3 = n$. If $m_1 + m_2 + m_3 = n$, and all $0 \le m_i \le n$, then, according to the polynomial law,

$$\mathbf{P}\{\nu_1 = m_1, \; \nu_2 = m_2, \; \nu_3 = m_3\}$$

$$= \frac{n!}{m_1! m_2! m_3!} p_1^{m_1} p_2^{m_2} p_3^{m_3}$$

One can easily verify that

$$\mathbf{P}\left\{ x \le \gamma_{(s)} < x + \Delta x \right\}$$

$$= \mathbf{P}\left\{ \nu_1 = s - 1, \; \nu_2 = 1, \; \nu_3 = n - s \right\} + O\left\{ (\Delta x)^2 \right\}$$

Then the density $p(x)$ of $\gamma_{(s)}$ can be computed:

$$p(x) = \lim_{\Delta x \to 0} \left(\mathbf{P}\{ x \le \gamma_{(s)} < x + \Delta x \} / \Delta x \right)$$

$$= n \binom{n-1}{s-1} x^{s-1} (1 - x)^{n-s}$$

Various random variables with densities (3.17) can be modeled using various order statistics. Assume that positive integers s and t are given and denote $n = s + t - 1$. Then the density (3.17) turns into a *beta-distribution*

$$p(x) = \frac{x^{s-1}(1 - x)^{t-1}}{B(s, \, t)} \text{ for } 0 < x < 1 \qquad (3.18)$$

where the beta-function

$$B(s, \, t) = \Gamma(s)\Gamma(t)/\Gamma(s + t)$$

$$= (s - 1)!(t - 1)!/(s + t - 1)!$$

Therefore, for modeling a beta-distributed random variable ξ, one has to select $s + t - 1$ random numbers $\gamma_1, \; \ldots, \; \gamma_{s+t-1}$, re-order them $\gamma_{(1)} \le \cdots \le \gamma_{(s+t-1)}$, and set $\xi = \gamma_{(s)}$.

Algorithms for modeling gamma- and beta-distributed random variables with non-integer parameters can be found in Rubinstein.[8]

Transformations
of the Type $\xi = g(\gamma_1, \ldots, \gamma_n, \ldots)$

This type of transformation includes transformations that depend on a random quantity of random numbers, so that the number of γ-s used for computing of a particular value of ξ may be different and even unrestricted. Of course the average number of random numbers required for computing one value of ξ must be finite.

The most frequently used transformations of this type are the *rejection techniques*. von Neumann's method for modeling continuous random variables is one of the earliest examples in which these techniques were applied.

The *rejection technique* utilizes a computation equation that has an acceptance rule. For example,

$$\xi = g(\gamma_1, \ldots, \gamma_m) \quad \text{if} \quad h(\gamma_1, \ldots, \gamma_m) > 0$$

For modeling the variable ξ, we select m random numbers $\gamma_1, \ldots, \gamma_m$. If the acceptance rule is met, that is, if $h(\gamma_1, \ldots, \gamma_m) > 0$, we compute $\xi = g(\gamma_1, \ldots, \gamma_m)$; otherwise, the set $\gamma_1, \ldots, \gamma_m$ is rejected, and new numbers $\gamma_1, \ldots, \gamma_m$ are selected.

The acceptance probability

$$e = \mathbf{P}\left\{ h(\gamma_1, \ldots, \gamma_m) > 0 \right\}$$

is called the *efficiency* of the technique.

If N random groups $(\gamma_1, \ldots, \gamma_m)$ are considered, then, on the average, only eN values of ξ are computed. Thus, eN values of ξ require mN random numbers, and, on the average, m/e random numbers have to be consumed in order to obtain one value of ξ. Clearly, when $e \to 0$ this modeling method becomes inefficient.

generalized von neumann's method

Assume that the density $p(x)$ of a random variable ξ can be represented as a product of the form

$$p(x) = p_1(x)f(x)$$

where $p_1(x)$ is the density of an auxiliary random variable η', and $f(x)$ is bounded: $f(x) \leq C$.

If we know how to model η' we can construct a method for modeling ξ:

1. Find η' and an independent random number γ; compute $\eta'' = C\gamma$.

2. If $\eta'' < f(\eta')$, set $\xi = \eta'$; otherwise reject η' and γ and repeat (1).

proof

The point (η', η''), in Figure 3.9, will be in the strip $(x, x + dx)$, with probability $p_1(x)\,dx$. Since η'' is uniformly distributed over the interval $0 < y < C$, the conditional probability of acceptance is $f(x)/C$. Accordingly, the probability of η' falling into $(x, x + dx)$ without being rejected is equal to the product $p_1(x)\,dx\,f(x)/C$, and is thus proportional to $p(x)\,dx$.

The efficiency of this method can be computed by summing the conditional acceptance probabilities:

$$e = \int_a^b \frac{f(x)}{C} p_1(x)\,dx = \frac{1}{C} \int_a^b p(x)\,dx = \frac{1}{C}$$

von Neumann's method is a special case of the last method, corresponding to the choice of a constant density $p_1(x) = 1/(b-a)$ and $f(x) = (b-a)p(x)$ in a finite interval $a < x < b$. The restriction $p(x) \leq M_0$ yields the bound $f(x) \leq (b-a)M_0$. Therefore, the efficiency of von Neumann's method is $e = \big((b-a)M_0\big)^{-1}$.

The same result can be obtained directly from Figure 1.11: since the random point $\Gamma(\eta', \eta'')$ is uniformly

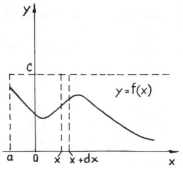

Fig. 3.9. Random points that appear above the curve $y = f(x)$ are rejected.

distributed over the rectangle $a < x < b$, $0 < y < M_0$, the acceptance probability is equal to the ratio of the area under the curve $y = p(x)$ to the area of the rectangle:

$$e = \int_a^b p(x)\,dx / \big((b-a)M_0\big) = \big((b-a)M_0\big)^{-1}$$

Example

Consider the density $p(x) = v(x)x^{-1/3}$ defined in $0 < x < 1$, with a bounded function $v(x) \le A$.

Let us choose an auxiliary density $p_1(x) = (2/3)x^{-1/3}$ that can be treated by the inverse functions method. Then the corresponding $f(x)$ is bounded:

$$f(x) = p(x)/p_1(x) = (3/2)v(x) \le (3/2)A$$

The acceptance rule $\eta'' < f(\eta')$ can be divided by $3/2$ and the algorithm for modeling ξ with density $p(x)$ thus acquires the form:

(1)   Select γ_1 and γ_2; compute $\eta' = (\gamma_1)^{3/2}$

(2)   If $A\gamma_2 < v(\eta')$, set $\xi = \eta'$; otherwise reject the pair γ_1, γ_2 and repeat (1).

on monte carlo algorithms

Time Consumption of a Monte Carlo Algorithm

Suppose that we are interested in a certain quantity m, and we have defined a random variable ξ so that its expectation $M\xi = m$, and its variance $D\xi$ is finite. Then we can select N independent values $\xi_1, \ldots, \xi_N$ of this variable, and form the estimate

$$m \approx \frac{1}{N} \sum_{j=1}^{N} \xi_j \qquad (3.19)$$

which is usually called a *Monte Carlo method for estimating* m. We have seen that the accuracy of the estimate (3.19) depends on $D\xi/N$ (see The General Scheme of the Monte Carlo Method, Chapter 1). However relation (3.19) still does not determine the computation algorithm; one requires an equation for modeling ξ by means of standard random numbers.

Let us specify the modeling equation

$$\xi = g(\gamma_1, \gamma_2, \ldots) \qquad (3.20)$$

Both relations (3.19) and (3.20) completely define the *Monte Carlo algorithm* for estimating m.

Let t denote the computer time expended in calculating a single value of ξ from (3.20). Then the total computing time of (3.19) is $T = Nt$.

Consider the expression for the probable error of estimate (3.19)

$$r_N = 0.6745\sqrt{D\xi/N}$$

and substitute $N = T/t$. Then

$$r_N = 0.6745\sqrt{t D\xi/T} \qquad (3.21)$$

The last formula shows that if the total computer time T is held fixed, then the probable error of a Monte Carlo algorithm depends on the product $t \cdot D\xi$, which is called the *time consumption* (or *laboriousness*) of the algorithm. The less laborious the algorithm, the smaller the probable error.

In most problems, both $\mathbf{D}\xi$ and t are estimated numerically from relatively small samples, however, sometimes this can be done analytically.

An example with splitting

Consider the problem of estimating the expectation $m = \mathbf{M}\theta$ of a random variable $\theta = f(\xi, \eta)$. Assume that subroutines for modeling ξ and η are available, as well as a subroutine for computing $f(\xi, \eta)$. Assume also that the variance $\mathbf{D}\theta$ is finite.

Clearly, we can model independent values $\theta_j = f(\xi_j, \eta_j)$ and use the estimate

$$m \approx \frac{1}{N} \sum_{j=1}^{N} \theta_j \qquad (3.22)$$

Let t_ξ and t_η be the computer time required to model ξ and η respectively, and let t_f be the computer time required for one calculation of $f(\xi, \eta)$. Then the time consumption of the algorithm (3.22) is $t \cdot \mathbf{D}\theta$, with $t = t_\xi + t_\eta + t_f$.

Now assume that the weak point of our computation algorithm is the modeling of ξ, or, more precisely, $t_\xi \gg t_\eta + t_f$. Then it may be worthwhile to consider averaged values

$$\theta^{(s)} = \frac{1}{s} \sum_{k=1}^{s} f(\xi, \eta_k)$$

rather than θ, so that each ξ is coupled with s independent values of η. The variance of $\theta^{(s)}$ can be expressed in the form

$$\mathbf{D}\theta^{(s)} = \mathbf{D}\theta(1 - r + rs)/s \qquad (3.23)$$

where r is a constant $0 \le r \le 1$. The computing time of $\theta^{(s)}$ is equal to

$$t^{(s)} = t_\xi + s(t_\eta + t_f)$$

(The constant r is the correlation coefficient of the random variables $f(\xi, \eta')$ and $f(\xi, \eta'')$, with η' and η'' being independent values of η. For obtaining $\mathbf{D}\theta^{(s)}$ the variance of a sum of random variables must be computed using the formula $\mathbf{D}\sum\limits_{k=1}^{s} f_k = \sum\limits_{k=1}^{s}\mathbf{D}f_k + 2\sum\limits_{1\le k<j\le s}\sum r(f_k, f_j)\sqrt{\mathbf{D}f_k\mathbf{D}f_j}.$)

We can write down the generalized estimate

$$m \approx \frac{1}{Ns}\sum_{j=1}^{N}\sum_{k=1}^{s} f(\xi_j, \eta_{jk}) \tag{3.24}$$

and the time consumption of the corresponding algorithm is

$$t^{(s)}\mathbf{D}\theta^{(s)} = \mathbf{D}\theta(1 - r + rs)(t_\eta + t_f + t_\xi/s)$$

The last expression has one minimum in the region $0 < s < \infty$, namely, at $s = s_*$, where

$$s_* = \sqrt{(1/r - 1)t_\xi/(t_\eta + t_f)}$$

Numerical example

Let $t_\xi \approx 100(t_\eta + t_f)$ and $r \approx 0.2$. Then $s_* \approx 20$. The time consumption of (3.24) with the optimal $s = s_*$ is 3.5 times less than the time consumption of (3.22) (Figure 3.10).

Turning from θ to $\theta^{(s)}$ is a very special case of a general Monte Carlo technique termed *splitting of a trajectory*. In our example the "trajectory" is one path from ξ to η.

Constructive Dimension of a Monte Carlo Algorithm

Consider an arbitrary algorithm of the type (3.19)–(3.20) for computing of an arbitrary quantity $m = \mathbf{M}\xi$. If Equation 3.20 depends on n variables only, that is, if

$$\xi = g(\gamma_1, \ldots, \gamma_n)$$

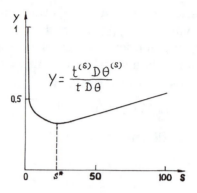

$$y = \frac{t^{(s)} D \theta^{(s)}}{t \, D \theta}$$

Fig. 3.10. The optimum value $s = s_*$.

we say that the *constructive dimension* of the algorithm (3.19)–(3.20) is n. For the sake of brevity we will write c.d.$= n$.

In other words, the c.d. is equal to the number of random numbers used for computing one value of ξ. More precisely, the c.d. is the maximum number of random numbers necessary for carrying out one trial. For example (see Computation of Neutron Transmission Through a Plate in Chapter 2), different neutrons undergo different numbers of scatterings; the amounts of random numbers used for computing one value of η are different also. Theoretically, the number of scatterings is unlimited, therefore, this algorithm has c.d. $= \infty$. However, if the number of scatterings that must be considered is *a priori* restricted, the c.d. of the algorithm is finite.

If an intermediate random variable ξ' (in the algorithm (3.19)–(3.20)) is modeled by an equation of type $\xi' = g'(\gamma_1, \ldots, \gamma_s)$, then the c.d. of the algorithm increases as s increases. Hence, the method of inverse functions is the best method if one wants to minimize the c.d. On the other hand, if ξ' is modeled by rejection technique, then the c.d. is infinite.

Arbitrary Monte Carlo algorithms with c.d.$= n$ can be interpreted as numerical integrations in n dimensions. Indeed, let Γ be an n-dimensional random point with Cartesian coordinates $(\gamma_1, \ldots, \gamma_n)$. The point Γ is uniformly distributed in the n-dimensional unit cube

$$\{0 \le x_1 \le 1, \ldots, 0 \le x_n \le 1\}$$

since its density is constant inside the cube:

$$p_\Gamma(x_1, \ldots, x_n) = p_{\gamma_1}(x_1) \ldots p_{\gamma_n}(x_n) = 1$$

In this case relation (3.20) is equivalent to $\xi = g(\Gamma)$.

Now consider the quantity m that we are estimating $m = \mathbf{M}g(\Gamma)$. Using a multidimensional analog of (1.18), we obtain

$$\mathbf{M}g(\Gamma) = \int_0^1 \ldots \int_0^1 g(x_1, \ldots, x_n) p_\Gamma(x_1, \ldots, x_n)\, dx_1 \ldots dx_n$$

$$= \int_0^1 \ldots \int_0^1 g(x_1, \ldots, x_n)\, dx_1 \ldots dx_n$$

Therefore, the relation (3.19) can be rewritten as

$$\int_0^1 \ldots \int_0^1 g(x_1, \ldots, x_n)\, dx_1 \ldots dx_n$$

$$\approx \frac{1}{N} \sum_{j=1}^N g(\Gamma_j) \qquad (3.25)$$

It is clear from (3.25) that our estimate is an approximation of an integral over the n-dimensional unit cube, derived by averaging the values of the integrand at independent random points $\Gamma_1, \ldots, \Gamma_N$ uniformly distributed in the cube.

Quasi-Monte Carlo Method

In 1916 H. Weyl found that infinite sequences of non-random points $Q_1, Q_2, \ldots, Q_j, \ldots$ exist, which have a property similar to (3.25): for an arbitrary Riemann-integrable function $g(x_1, \ldots, x_n)$

$$\int\limits_0^1 \cdots \int\limits_0^1 g(x_1, \ldots, x_n)\, dx_1 \ldots dx_n$$

$$= \lim_{N \to \infty} \frac{1}{N} \sum_{j=1}^N g(Q_j) \qquad (3.26)$$

Such sequences are said to be *uniformly distributed in the number-theoretical sense.*

Number-theoreticians were initially interested only in the asymptotic behavior of these sequences. However, sequences intended for numerical computations must satisfy not only (3.26), but several additional requirements also (see Sobol'):[3] The uniformity of distribution should be optimal as $N \to \infty$; the uniformity of distribution of initial points $Q_1, \ldots, Q_N$ should be observed for fairly small N; and formulas for computing these points should be simple.

The first sequences satisfying to a certain extent all of these requirements were the LP_τ-sequences constructed in 1966. Programs for generating these sequences can be found in Bratley and Fox (FORTRAN-77),[9] or in Sobol' et al (FORTRAN-77 and C).[10]

Comparing (3.25) with (3.26), one can conclude that if random points Γ_j in (3.25) are replaced by points Q_j, then the averages converge. In practice, as a substitute for the value $\xi_j = g(\Gamma_j)$, one has to use the value $\xi_j = g(Q_j)$, or, in other words, the jth trial should be carried out using Cartesian coordinates $(q_{j1}, \ldots, q_{jn})$ of the point Q_j rather than random numbers $\gamma_1, \ldots, \gamma_n$.

Points $Q_1, Q_2, \ldots$ are often called *quasi-random.* When standard statistical tests are applied for verifying the distribution of these points in the n-dimensional cube, the results are "too good" for random points. (Of course these points lack many other properties of random points as well.)

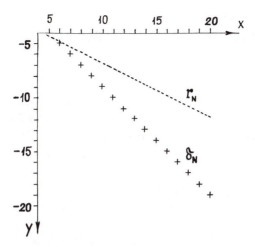

Fig. 3.11. Computation errors δ_N and probable errors r_N (logarithmic scale: $x = \log_2 N$, $y = \log_2 \delta_N$ or $\log_2 r_N$).

The main reason in favor of quasi-random points is the possible increase in the rate of convergence: while the approximation error of (3.25) for large N is decreasing as $1/\sqrt{N}$, the convergence rate of (3.26) can sometimes even be $1/N$. Thus, without changing the computation algorithm (3.19)–(3.20), but merely by replacing the random numbers with coordinates of quasi-random points, we can sometimes considerably improve our results.

According to Sobol' et al,[10] if all the variables $x_1, \ldots, x_n$ in $g(x_1, \ldots, x_n)$ are equally important, and n is large (say, $n > 15$), then there is no advantage in switching to quasi-Monte Carlo methods. However, if the dependence of $g(x_1, \ldots, x_n)$ on x_i decreases as i is increased (in other words, the initial coordinates are the leading ones), one can expect a considerable benefit from employing quasi-Monte Carlo methods, even if n is large (say, several tens).

Numerical example

Consider the random variable

$$\xi = \frac{1}{10!}(1 + 2\gamma_1)(2 + 2\gamma_2) \ldots (9 + 2\gamma_9)$$

Its expectation is $M\xi = 1$, and its variance $D\xi = 0.196$. If $M\xi$ were estimated by the Monte Carlo method (3.19) the probable error would be $r_N = 0.30N^{-1/2}$.

The corresponding quasi-Monte Carlo method

$$M\xi \approx I_N = \frac{1}{N}\sum_{j=1}^{N}\frac{1 + 2q_{j1}}{2}\frac{2 + 2q_{j2}}{3} \cdots \frac{9 + 2q_{j9}}{10}$$

where $(q_{j1}, \ldots, q_{j9})$ is the jth point of a 9-dimensional LP_τ-sequence is realized. The computational errors $\delta_N = |I_N - 1|$ are shown in Figure 3.11; the scale is logarithmic: $x = \log_2 N$, $y = \log_2 \delta_N$; the dotted line is $y = \log_2 r_N$.

One can see that $\delta_N \approx 2/N$.

notes

references

1. Metropolis, N. and Ulam, S., The Monte Carlo method, *J. Am. Stat. Assoc.*, 44, N 247, 335–341, 1949.

2. RAND Corporation, *A million random digits with 100,000 normal deviates*, Free Press, Glencoe, 1955.

3. Sobol', I. M., *Chislennye Metody Monte Karlo (Numerical Monte Carlo methods)*, Nauka, Moscow, 1973.

4. Pozdnyakov, L. A., Sobol', I. M., and Syunyaev, R. A., Comptonization and the shaping of X-ray source spectra: Monte Carlo calculations, in *Soviet Scientific Reviews, Sect. E: Astrophys. and Space Phys. Reviews*, Harwood Academic, New York, 2, 189–331, 1983.

5. Forsythe, G. E., Malcolm, M. A., and Moler, C. B., *Computer Methods for Mathematical Computations*, Prentice-Hall, Englewood Cliffs, NJ, 1977.

6. Wichmann, B. A. and Hill, I. D., An efficient and portable pseudo-random number generator, *Appl. Stat.*, 31, N 2, 188–190, 1982.

7. Levitan, Yu. L. and Sobol', I. M., On a pseudo-random number generator for personal computers, *Matem. Modelirovanie* (*Math. Modeling*), 2, N 8, 119–126, 1990.

8. Rubinstein, R. Y., *Simulation and the Monte Carlo Method*, John Wiley & Sons, New York, 1986.

9. Bratley, P. and Fox, B. L., Implementing Sobol's quasirandom sequence generator, *ACM Trans. Math. Software*, 14, N 1, 88–100, 1988.

10. Sobol', I. M., Turchaninov, V. I., Levitan, Yu. L., and Shukhman, B. V., *Quasirandom sequence generators*, Keldysh Institute of Applied Mathematics, Moscow, 1992.

index